AF597363

Electronic Circuit Guidebook
Volume 2:
IC Timers

Electronic Circuit Guidebook
Volume 2:
IC Timers

Written By
Joseph J. Carr

A Division of Howard W. Sams & Company
A Bell Atlantic Company
Indianapolis, IN

PROMPT© Publications is an imprint of Howard W. Sams & Company, A Bell Atlantic Company, 2647 Waterfront Parkway, E. Dr., Indianapolis, IN 46214-2041.

International Standard Book Number: 0-7906-1106-6
Library of Congress Card Catalog Number: 97-65787

Acquisitions Editor: Candace M. Hall
Editor: Loretta L. Leisure
Assistant Editors: Pat Brady, Natalie F. Harris
Layout Design: Loretta L. Leisure
Typesetting: Loretta L. Leisure
Cover Design: Christy Pierce
Graphics Conversion: Jason Coats, Terry Varvel
Illustrations and Other Materials: Courtesy of the Author

PRINTED IN THE UNITED STATES OF AMERICA

9 8 7 6 5 4 3 2 1

Contents

Preface

Timer circuits used to be a lot of trouble to build and tame for several reasons. One major reason was the fact that DC power supply variations would cause a frequency shift or slow drift. Modern timers are designed to permit a wide variation in the DC supply voltage without affecting the output frequency or single-pulse duration. Another reason is that the equations determining the frequency or timing were little more than approximate, despite the appearance of great precision in the textbooks. Modern timers will work very closely to the way the textbooks say they do.

There are a number of modern timers on the market. Some of them are old favorites. The LM-555 device, for example, has been around since the mid-1970s, but is adaptable to so many different purposes that it maintains a large following. Popularity over long time periods is not something integrated circuits see!

The first part of this book is organized to demonstrate the theory of how the timers work, with application examples coming later. Feel free to modify these circuits to your own use. In fact, I hope this book teaches you enough that you not only can rework and modify my circuits, but design a few of your own.

Connections

The author can be reached through the publisher, or at PO Box 1099, Falls Church, VA, 22041, or via e-mail at carrjj@aol.com.

Joseph J. Carr

Part I

Overview of IC Timers

Introduction to RC Timer Circuits

TTL and CMOS Digital IC Timers

The LM-555 RC Timer IC Device

555 Monostable Operation

555 Astable Multivibrator Circuits

Other IC Timers

Operational Amplifier Timer Circuits

Retriggerable Timers

Long Duration Timers

Chapter 1

Introduction to RC Timer Circuits

Timer circuits come in two basic varieties (and several sub-varieties): monostable and astable multivibrators. The monostable multivibrator (MMV), also called the one-shot circuit, produces a single output pulse of duration *T* for every valid trigger input pulse. The astable multivibrator (AMV), on the other hand, produces a train of pulses or square waves depending on the design.

The classes of timer circuits that we will discuss are those that are controlled by a resistor-capacitor (RC) network. Others surely exist; but those in which an RC circuit controlls the pulse duration or pulse repetition rate are of interest here.

We will take a look at both specialist integrated circuit MVs (e.g., LM-555), digital logic family MVs and operational amplifier MVs.

Monostable Multivibrators (MMV or One-Shot)

Figure 1-1a shows the block diagram to an MMV circuit. When a trigger pulse is received on the TRIG IN input, a single output pulse is generated. These circuits are said to be quasistable because they have two states: a stable dormant state and an unstable state.

Normally, the MMV is in the dormant state, which most usually means that the output signal is at 0V. But the stable or dormant state could just as easily be defined as some particular voltage. When the trigger pulse is received, the output state changes to the unstable state, where it remains for a specified period of time, *T*. When *T* expires, then the output snaps back to the dormant state where it remains until another trigger pulse is received.

For most MMV circuits there is a rule in effect regarding the relationship of the trigger pulse to the output pulse. If *t* is the duration of the trigger pulse, and *T* is the duration of the output pulse, then *T* must be very much larger than *t*, i.e., $T >> t$.

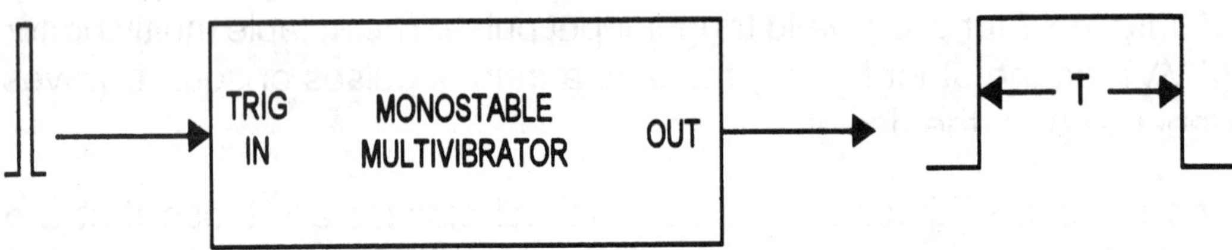

Figure 1-1a. Operation of the monostable multivibrator uses a trigger pulse to create an output pulse

So, of what use is a circuit that produces output pulses on a one-for-one basis with input pulses? One application is shown in **Figure 1-1b**. In this circuit, the input waveform is a pulse, but it is either noisy or distorted (possibly both). The input pulse is reconstituted by generating a pulse of constant amplitude and duration. The output pulse is more useful than the uncleaned version (indeed, in some circuits the trigger input pulse won't work at all).

Figure 1-1b. One application is to clean up a noisy or distorted input pulse.

Another application is to stretch pulses. A source might have, say, 1 mS pulses but the requirement is for 50 mS pulses. The 1 mS pulse can be used to generate a 50 mS pulse by triggering a 50 mS MMV circuit.

The MMV can also be used to provide a delay in a pulse. In that type of application, there will be two monostable multivibrators, one feeding the other. Suppose there is a trigger pulse of 2 mS, the first multivibrator (MMV1) is set to 25 mS, and the second (MMV2) to 2 mS. The reconstituted 2 mS pulse output from MMV2 is essentially a delayed version of the trigger input to MMV1, but delated by the 25 mS duration of the MMV1 output pulse.

Input/Output Polarities. The input trigger pulse and output pulse shown in **Figure 1-1a** are both positive-going. That is, they are normally at 0V, and then snap to some positive voltage when active. But that is not an essential requirement. In **Figure 1-2** we see two permissible trigger pulses, one that is positive-going and one that is negative-going. The output pulses can also be either positive-going or negative-going as needed.

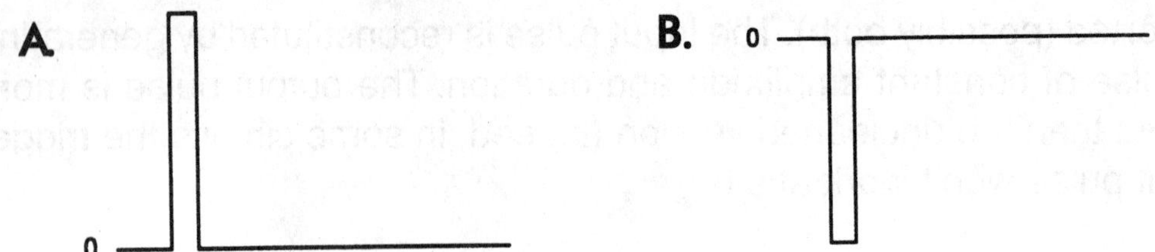

Figure 1-2. A monostable multivibrator can be triggered by either a) positive-going or b) negative-going pulses.

Monostable Multivibrator Trigger Categories

Figure 1-3a shows the conventional relationship between the trigger pulse and output pulse. The output is normally LOW (e.g., 0V), and can remain in that state indefinitely. When the trigger pulse is received, however, it snaps HIGH to some positive voltage for a specified time, *T*. When *T* expires, it drops back to the LOW state.

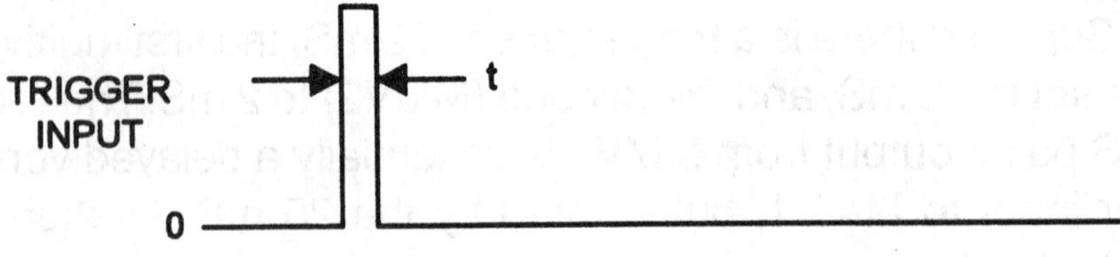

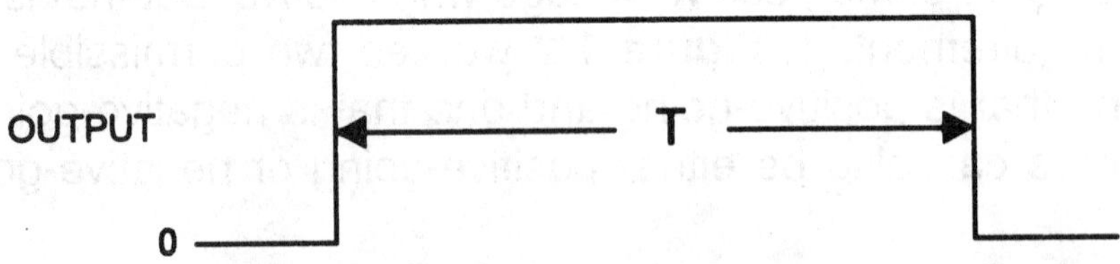

Figure 1-3a. Normal operation of the monostable multivibrator is to permit one and only one output pulse per trigger pulse.

One of the advantages of the MMV circuit is that it does not normally respond to additional input trigger pulses until the duration has expired. In **Figure 1-3b**, the first output pulse is triggered at time T1. Additional trigger pulses are received at times T2 and T3, but these do not affect the MMV output state at all. At time T4, however, the next trigger pulse is received and it is after the time-out of the output. Therefore, the trigger pulse at time T4 triggers another output pulse.

One use for this attribute of the MMV circuit is contact debouncing. Mechanical contacts used in switches and electromechanical relays don't make a clean closure, but rather they bounce a few times. As a result, a signal that is generated by opening or closing a switch could send a short train of pulses to the following circuit. The keyboard on the computer used to type this manuscript has a 5 mS delay to provide key debouncing. If the switch is used to trigger a one-shot, which has a 5 to 10 mS duration, then the contacts are automatically debounced.

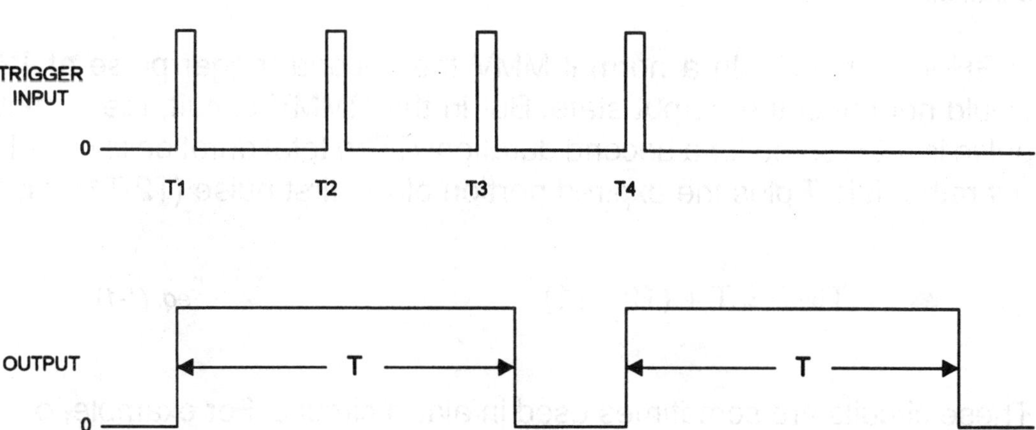

Figure 1-3b. Additional trigger pulses (T2 and T3) are ignored until after the duration T is expired (T4).

The timing waveforms for a retriggerable monostable multivibrator (RMMV) are shown in **Figure 1-3c**. These circuits are specially designed to permit retriggering. In the version of **Figure 1-3c** an initial trigger pulse is received at time T1, and it causes the output to snap

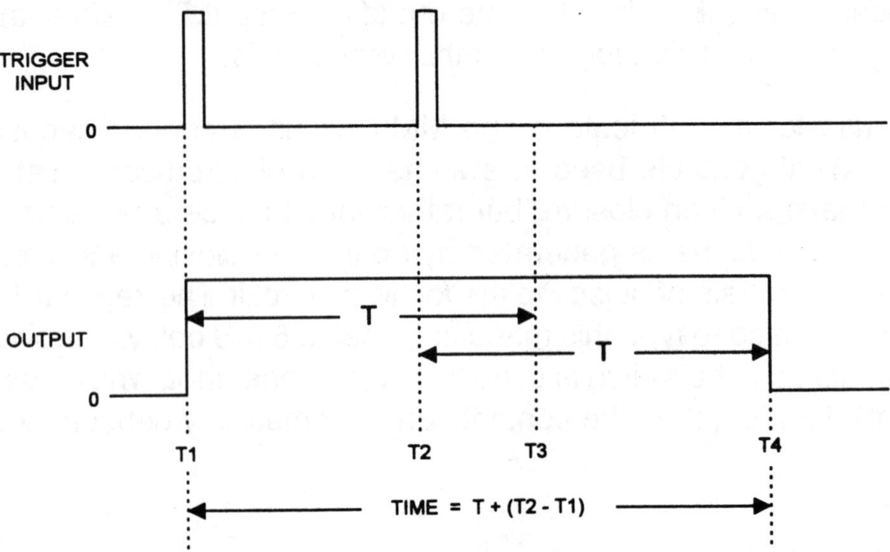

Figure 1-3c. A retriggerable monostable multivibrator will accept additional trigger pulses, adding one more duration* T *to the expired portion of the previous pulse duration.

HIGH for a time *T*. In a normal MMV the second trigger pulse at T2 would not affect the output state. But in the RMMV circuit, the output pulse is retriggered for a second duration *T*. The total duration is not 2T, but rather it is *T* plus the expired portion of the first pulse (T2-T1), or,

$$\text{TIME} = T + (T2 - T1) \qquad \textit{eq. (1-1)}$$

These circuits are sometimes used in alarm circuits. For example, one company produces a medical respirator alarm. The signal from a transducer in the patient's exhalation tube senses small pressure wave variations not dissimilar to the input waveform shown previously in **Figure 1-1b**. These pulses are cleaned up in a circuit such as a zero-crossing detector, a Schmitt trigger or a short-duration MMV circuit. The cleaned pulse is then used to trigger the RMMV. Let's say that the RMMV duration was 10 seconds. As long as pulses are received from the breathing sensor at intervals of less than 10 seconds apart, the alarm will remain silent. But if the 10 seconds expires because no respiration signal was received, an alarm will sound.

The same concept was once used in cassette tape players to provide either a direction reversal or an automatic shut-off capability. A small wheel turned with the take-up hub of the cassette. It alternated copper/insulator, so a wiper brush against the wheel would alternately make connection and then not. The result was a string of pulses (about 2/sec). These pulses were used to retrigger a 5 second RMMV. When the wheel stopped turning, the RMMV would time out and the circuitry would take the correct action to reverse direction or turn off, as needed.

Review of Resistor-Capacitor Networks

This section is provided as a brief review of RC network DC theory. Consider **Figure 1-4a**. Assuming that the initial condition is as shown, switch S1 is in position-A and is thus open-circuited. There is initially no

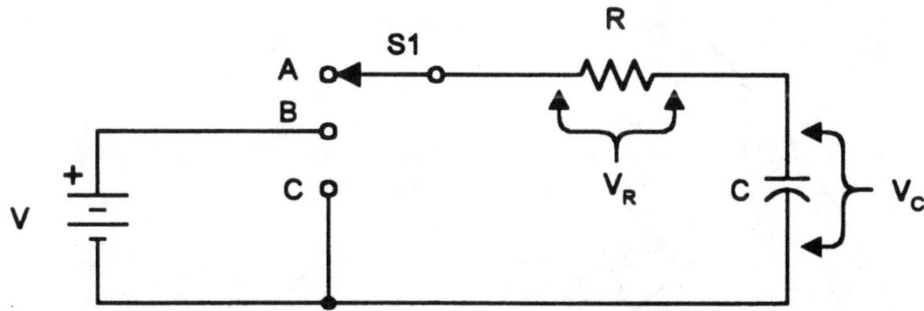

Figure 1-4a. A simple RC test circuit will permit a capacitor to charge, and then discharge, depending on switch (S1) setting.

electrical charge stored in capacitor C (i.e., $V_c = 0$). If switch S1 is moved to position-B, however, voltage V is applied to the RC network. The capacitor begins to charge with current from the battery, and V_c begins to rise towards V (see **Figure 1-4b**). The instantaneous capacitor voltage is found from:

$$V_c = V[1 - e^{-T/RC}] \qquad \text{eq. (1-2)}$$

Where:

V_c is the capacitor charge potential in volts (V)

V is the applied potential from the source in volts (V)

T is the elapsed time after charging begins in seconds (S)

R is the resistance in ohms (Ω)

C is the capacitance in farads (F)

e is the logarithm constant 2.718

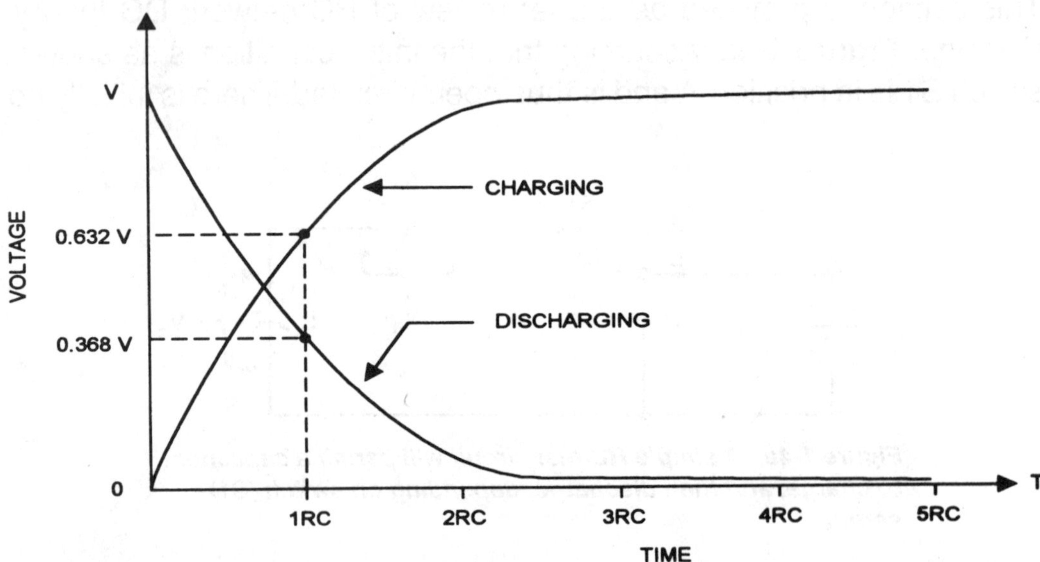

Figure 1-4b. Normal charge and discharge waveforms.

The product RC is called the RC time-constant of the network, and is sometimes abbreviated *t.* If *R* is in ohms, and *C* is in farads, then the product *RC* is in seconds. The capacito1=r voltage rises to approximately 63.2 percent of the final value after 1RC, 86 percent after 2RC and >99 percent after 5RC. A capacitor in an RC network is considered fully charged (by definition) after five time-constants.

If switch S1 in **Figure 1-4a** is next set to position-C, the capacitor will begin to discharge through the resistor. In the discharge condition:

$$V_c = V e^{-T/RC} \qquad \text{eq. (1-3)}$$

Voltage V_c drops to 36.8 percent of the full charge level after one time-constant (1RC), and to very nearly zero after 5RC.

Next consider **Figure 1-4c**. This graph represents a situation commonly encountered in waveform generator circuits. In this graph the capacitor is required to charge from some initial condition (V_{C2}), which may or

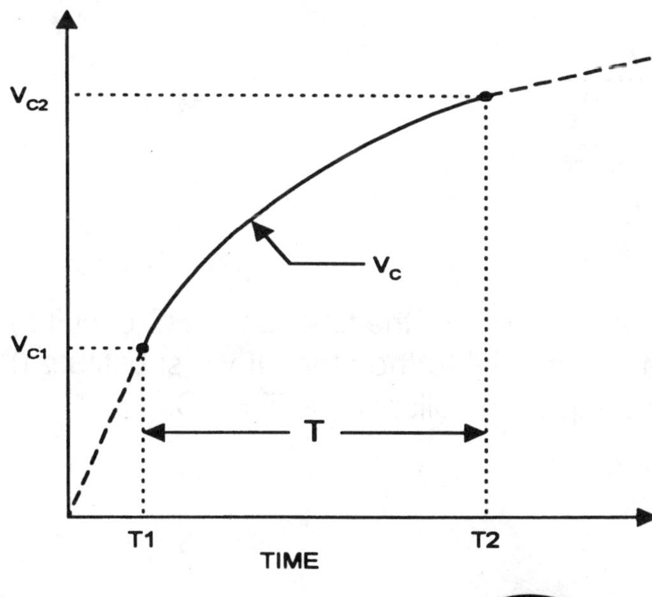

Figure 1-4c. A portion of the waveform seen in some cases where the initial and final voltages of the capacitor are not 0-V and V+.

may not be the fully charged "5RC" point, in a specified time interval, T. The question asked is: "What RC time constant will force V_{C1} to rise to V_{C2} in time *T*? Assuming that $V_{C1} < V_{C2} < V$:

- $$V - V_{C2} = (V - V_{C1})\, e^{-T/RC}$$ eq. (1-4)

- $$\frac{V - V_{C2}}{V - V_{C1}} = e^{-T/RC}$$ eq. (1-5)

- $$\mathrm{LN}\left[\frac{V - V_{C2}}{V - V_{C1}}\right] = \frac{-T}{RC}$$ eq. (1-6)

or, rearranging terms:

- $$RC = \frac{-T}{\mathrm{LN}\left[\frac{V - V_{C2}}{V - V_{C1}}\right]}$$ eq. (1-7)

Equation 1-7 is the general expression for the time for an RC circuit to charge or discharge from one potential to the other. If we substitute 0 for V1, and V for V2, then the equation collapses to T = RC.

Example:

An RC network is connected to a +12 VDC source. What RC product will permit voltage V_c to rise from +1 VDC to +4 VDC in 200 mS when V = +12 VDC, V_{C2} = +4 VDC and V_{C1} = +1 VDC?

Solution:

- $$LN\left[\frac{V - V_{C2}}{V - V_{C1}}\right] = \frac{-T}{RC}$$ *eq. (1-8)*

- $$RC = \frac{-\left[200\ ms \_ \frac{1\ s}{1000\ ms}\right]}{LN\left[\frac{12 - 4}{12 - 1}\right]}$$ *eq. (1-9)*

- $$RC = \frac{-0.200\ \text{sec}}{LN\left[\frac{8}{11}\right]} = \frac{-0.200\ \text{sec}}{LN\ (0.727)}$$ *eq. (1-10)*

- $$RC = (-0.200\ \text{sec})(-0.319) = 0.0627\ \text{sec}$$ *eq. (1-11)*

Equation 1-7 can be used to derive the timing or frequency setting equations of many different RC-based waveform generator circuits. The key voltage levels will, most often, be trip points or critical values set by the design of the circuit.

You will see the results of **Equation 1-7** brought to bear in some of the circuits described in this book. Indeed, all of the RC network theory that is presented in this chapter is relevant to the rest of the book, whether astable or monostable multivibrators are discussed.

Chapter 2

TTL and CMOS Digital IC Timers

In this chapter we will take a look at the monostable multivibrator chips available in the TTL and CMOS digital logic families. These devices are readily available from a large number of different sources.

Transistor-Transistor-Logic (TTL)

The *Transistor-Transistor-Logic* (TTL or T^2L) logic family emerged in the mid to late 1960s, and was one of the earliest attempts to standardize logic devices used in digital circuits. It was also a lot faster in operation than earlier logic families (RTL and DTL), as well as discrete logic circuits. These devices are given type numbers in the 74xx and 74xxx series, where "xx" and "xxx" denote the specific type number and the "74" denotes TTL. Sometimes you will also see 54xx and 54xxx type numbers. These devices are nothing more than the equivalent 74xx/74xxx devices, but in military packaging and temperature range. For example, a 54121 is nothing more than a 74121 in uniform.

TTL devices operate from a +5 VDC power supply and ground. The actual supply voltage should be held to the range +4.75 to +5.20 volts. Practical experience indicates that a narrower range, i.e., +4.9 to +5.2 volts, is prudent for complex function TTL chips such as the monostable multivibrators discussed in this chapter. Performance is "unspecified" (read, "beats the heck outa me") at voltages below the lower limit, and damage may occur at voltages above the maximum limit. The voltage supply should be regulated. In most cases, standard three-terminal IC voltage regulators with a +5 volt output will suffice (e.g., 7805, LM-340T-05, LM-340K-05 and LM-323).

The TTL devices operate at speeds in the range of 15 to 25 MHz, although a few 50 MHz devices are available. There are also sub-families of TTLs that operate at faster speeds.

The speed of the TTL devices creates a relatively high current demand. A typical device requires between 15 and 30 mA, with some devices asking for as high as 50 mA. If 22 mA is taken as the average value, then the current requirements for a large circuit rapidly add up; about 45 devices will require a 1 ampere DC power supply. Circuits that use more than a few TTL devices will generate a lot of heat and draw a lot of current. It might be wise in that case to consider whether or not TTL devices are the best selection. Other logic families (e.g., CMOS) draw considerably less current.

Figure 2-1 shows the typical TTL output stage (TTL1). The stages prior to this stage differ according to the type of device. Also shown is the input circuit of the following device (TTL2). The output stage in the LOW condition acts as a current sink, while the input serves as a current source. In the regular TTL family an input source is 1.6 mA. The output stage is usually designed to drive up to ten inputs, so it will sink 16 mA.

The fan-in and fan-out ratings of the device refer to the numbers of inputs supported. For example, a fan-in of 1 indicates a single 1.6 mA current source on regular TTL. The fan-out number is the number of TTL inputs that can be driven by an output. Thus, a standard output that will drive ten TTL inputs has a fan-out of 10.

The circuit in **Figure 2-1** is sometimes called a totem pole amplifier. It is characterized by having two identical transistors (in this case Q1 and Q2, which are NPN devices) across the DC power supply. There is a diode in series with the emitter of Q1 that blocks voltage applied to the output from affecting Q1.

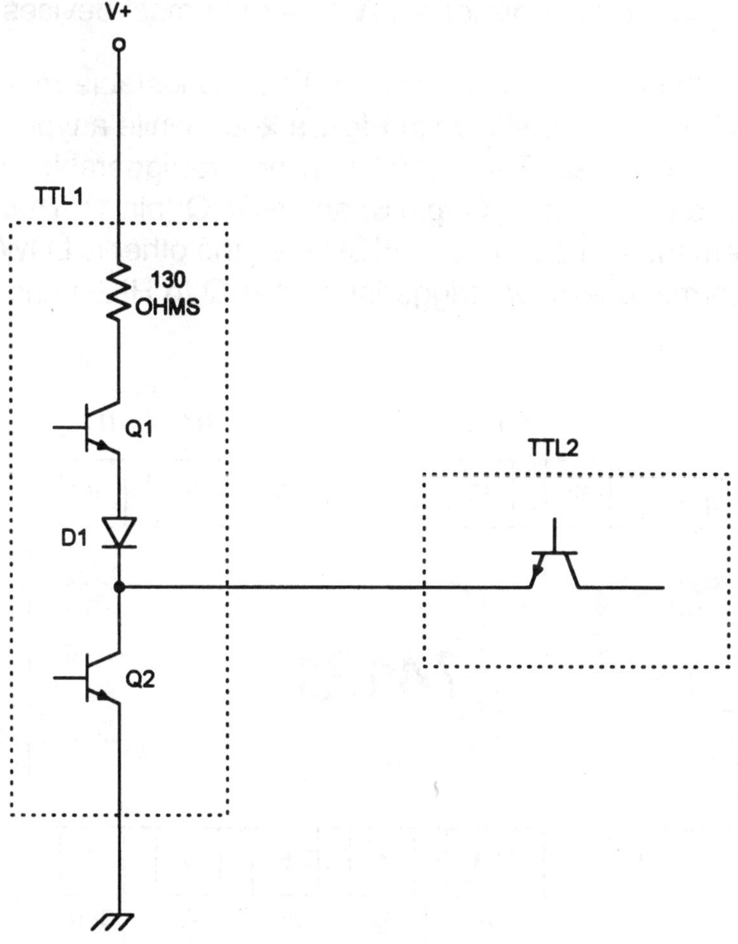

Figure 2-1. TTL output (TTL1) and input (TTL2) stages.

The input and output logic levels of TTL devices are related to the DC power supply voltages. The LOW condition (binary-0) is ground potential, or 0V. In actual practice, the device will see or produce the LOW condition whenever the voltage is between 0V and 0.8V. The HIGH condition is recognized when the voltage is between +2.4V and +5.2V. When the output of TTL1 is in the LOW condition, transistor Q1 is turned off, and Q2 is turned on hard. This condition puts the collector of Q2 at or near ground potential, so sinks the current from TTL2. When the output of TTL1 is HIGH, on the other hand, the reverse is true: Q1 is turned on hard and Q2 is turned off. As a result, the V+ supply, less the drop across the internal 130 ohm resistance, appears at the output. This voltage is on the order of +2.4V to +3V in most devices.

Figures 2-2 through 2-4 show several TTL monostable multivibrators. The 74121 pin-outs are shown in **Figure 2-2a**, while a typical circuit is shown in **Figure 2-2b**. The 74121 is a non-retriggerable MMV. Note that there are two outputs, Q (pin 6) and NOT-Q (pin 1). These outputs are complementary, i.e., if one is HIGH then the other is LOW and vice-versa. In normal operation, triggering forces Q to HIGH and NOT-Q to LOW.

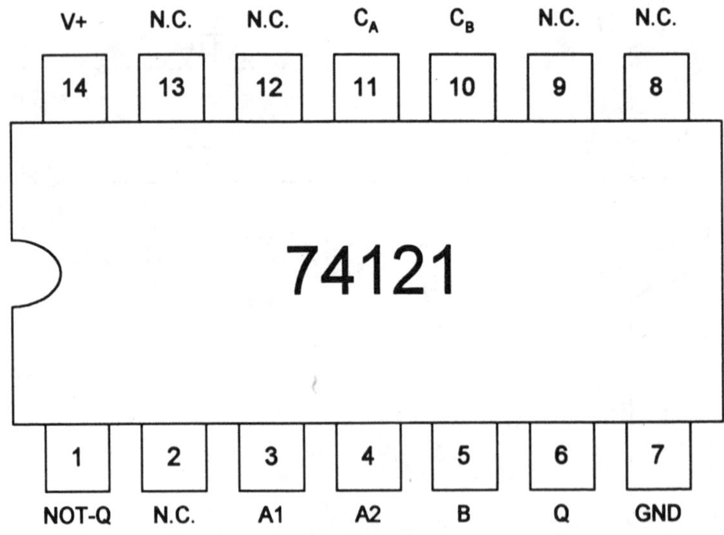

Figure 2-2a. The 74121 MMV circuit pin-outs.

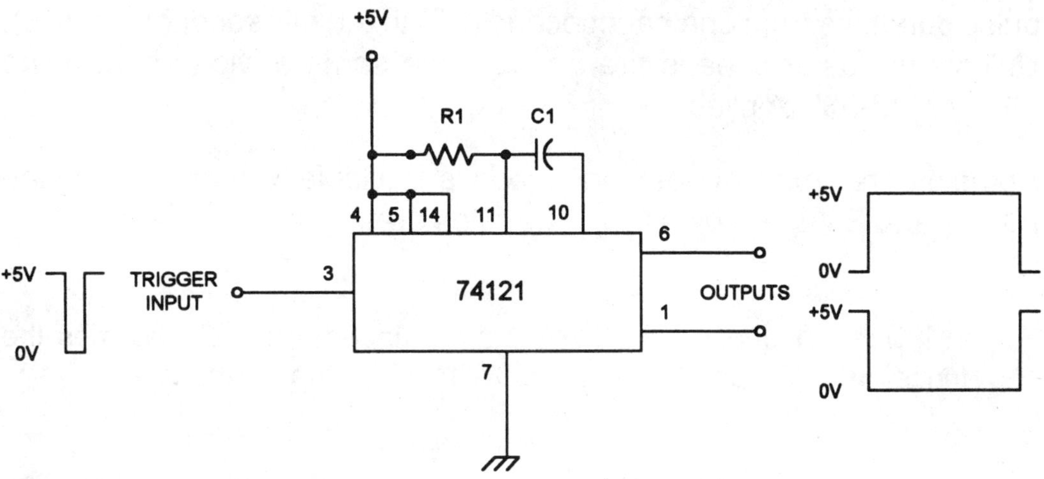

Figure 2-2b. The 74121 MMV circuit.

The timing of the output pulse duration is determined by R1 and C1 (**Figure 2-2b**) according to the equation:

- $T = 0.69 \, R1 \, C1$ *eq. (2-1)*

Where:

T is the output pulse duration in seconds (S)

R1 is expressed in ohms (Ω)

C1 is expressed in farads (F)

In practice it is necessary to keep R1 between 2 kohms and 40 kohms, and C1 greater than 10 pF. For timers requiring capacitors larger than about 0.001 μF (1,000 pF), it would probably be better to use a CMOS timer such as the 4528 or an LM-555 operated in the monostable mode. The 74121 device (as well as the 74122 and 74123 to follow) produce pulse durations from one nanosecond (nS) to ten milliseconds (10 mS). Unless the faster speeds are needed, the same advice about using 4528 or LM-555 applies.

Some programming of operating mode is available by manipulating the A1, A2 and B inputs (pins 3-5). The rules are:

1. If both A1 and A2 are LOW (i.e. grounded), then B becomes the trigger input. It operates on the LOW-to-HIGH input excursion.

2. If both A1 and B are HIGH, then A2 is the trigger input. It will trigger on the HIGH-to-LOW input excursion.

3. If both A2 and B are HIGH, then A1 is the trigger and the action occurs on the HIGH-to-LOW input excursion.

Other combinations of A1, A2 and B can be tried, but operation may be either inhibited or unpredictable. All trigger inputs must be properly terminated either HIGH or LOW (according to the rules above) or noise may cause triggering.

The 74122 device is shown in **Figure 2-3a** and **Figure 2-3b**. This device is similar to the 74121, but is retriggerable. Although the pins differ, the timing of the 74122 is similar to that of 74121 so that discussion will not be repeated.

There are five inputs that control triggering and operation: A1, A2, B1, B2 and CLR (clear). The rules are as follows:

1. The CLR input is normally kept HIGH (for example, tying it to +5 VDC). If CLR is brought LOW, then the input triggering is inhibited, Q is forced LOW and NOT-Q is forced HIGH.

2. A1, A2 and B2 HIGH: B1 is triggered on a LOW-to-HIGH input excursion.

3. A1, B1 and B2 HIGH: A2 is triggered on a HIGH-to-LOW input excursion.

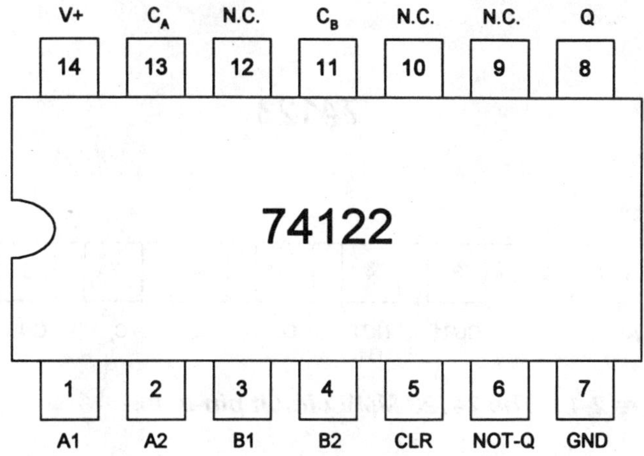

Figure 2-3a. The 74122 MMV circuit pin-outs.

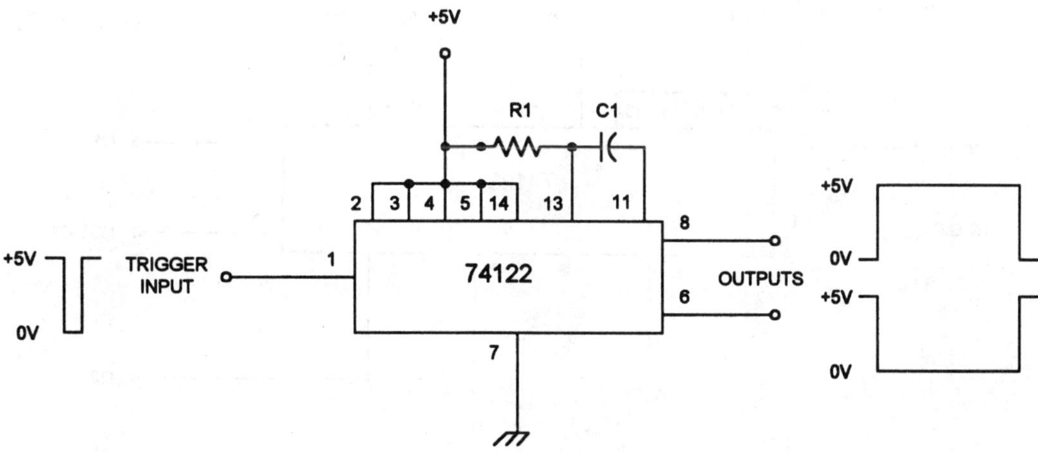

Figure 2-3b. The 74122 MMV circuit.

The 74123 device is shown in **Figure 2-4a** and **Figure 2-4b**. It is a dual monostable multivibrator. The CLR1 and CLR2 inputs operate the same way as CLR on the 74122.

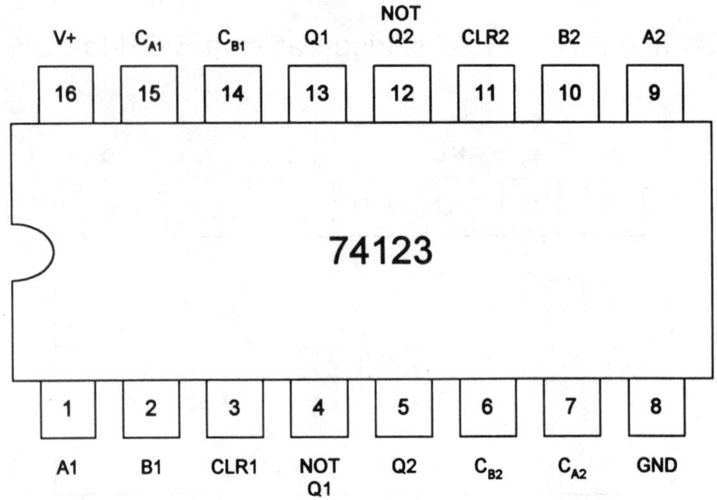

Figure 2-4a. The 74123 MMV circuit pin-outs.

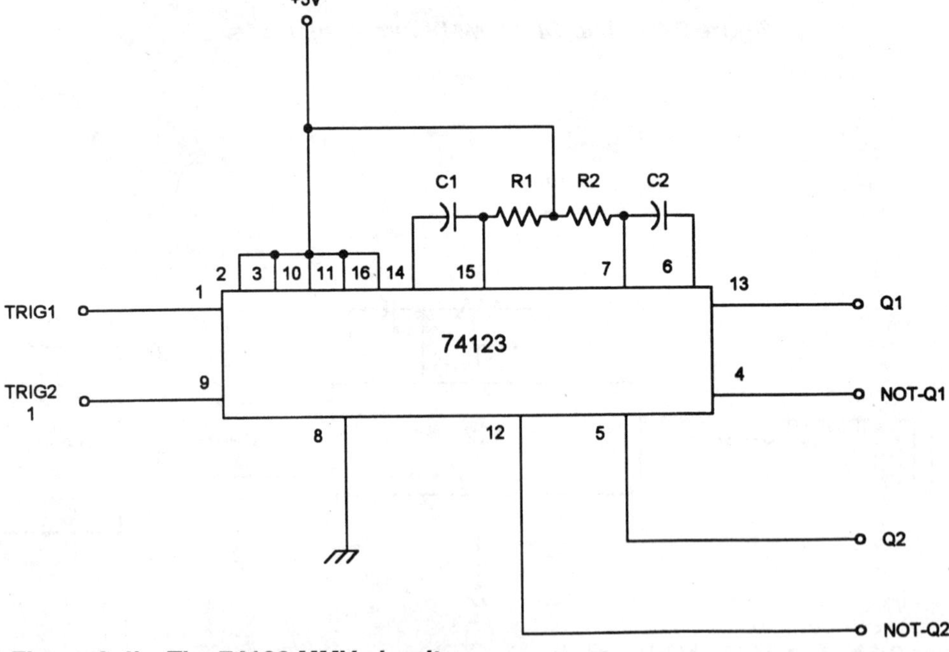

Figure 2-4b. The 74123 MMV circuit.

The following rules are applied to the 74123:

1. A1 is LOW: B1 is the trigger input, and will toggle the outputs when a LOW-to-HIGH input excursion occurs.

2. B1 is HIGH: A1 is the trigger input, and will toggle the outputs when a HIGH-to-LOW input excursion occurs.

Complementary Metal Oxide Semiconductor (CMOS)

When the CMOS logic family was introduced some years ago, pundits of the electronics industry prophesied that it would drive other logic families (especially TTL) off the market. That has not happened, but the CMOS family is clearly the family of choice for a large number of applications.

CMOS devices differ from other digital ICs because they use *metal oxide semiconductor field effect transistors* (MOSFETs) rather than bipolar NPN devices. It is a saturated logic family (like TTL), but operates at very much lower current levels than TTL devices.

For purposes of discussion we will consider the CMOS inverter as our example. An inverter complements (inverts) the input signal; i.e. a HIGH input makes a LOW output, and a LOW input makes a HIGH output. It takes two different MOSFETs to make an CMOS inverter circuit: *p-channel* (Q2) and *n-channel* (Q1). The p-channel device turns on when its gate potential is zero with respect to the source, while the n-channel turns on when the gate is positive with respect to the source.

Figure 2-5 shows a basic CMOS inverter circuit that is designed with a single n-channel and single p-channel device. This circuit also serves as the output stage in more complex function blocks. MOSFET transis-

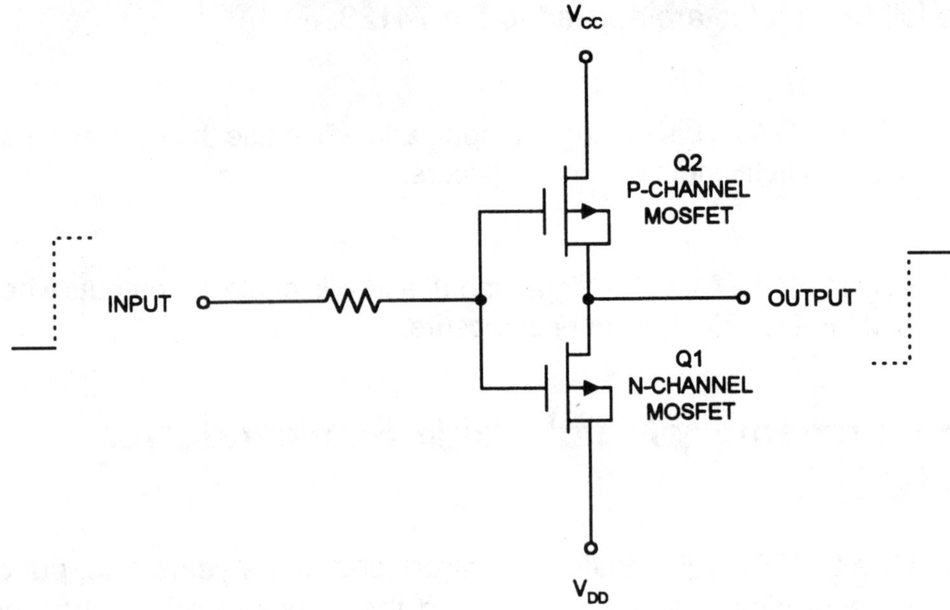

Figure 2-5. CMOS inverter circuit.

tors Q1 and Q2 are connected in series across the DC power supply, with the output taken at the junction of the two transistors. The input is created by connecting the gates of Q1 and Q2 in parallel, so the same input potential is applied to both simultaneously.

When the input is LOW (0V or some negative potential, depending on where V_{DD} is grounded or connected to a negative DC voltage), transistor Q1 is turned off and Q2 is turned on. This condition allows the V_{CC} potential to appear at the output through the channel resistance of Q2. Only a very small resistance is in series with the V_{CC} terminal and output terminal.

When the input signal goes HIGH, on the other hand, exactly the opposite situation occurs: Q1 is turned on and Q2 is turned off. In this condition there is a high resistance between the output terminal and V_{CC}, and a low resistance between the output terminal and ground or the V_{DD} power supply (whichever is used).

A model of the CMOS circuit action is shown in **Figure 2-6**, using resistors and the DC power supply. In this version, the V_{DD} is set to 0V so the source of Q1 is grounded. The resistances in Figure 2-6 represent the output conditions of Q1 and Q2 under two different circumstances: output LOW and output HIGH.

In **Figure 2-6a** the input is LOW, in which case the output is HIGH. Under this condition Q1 is turned off and Q2 is turned on. This makes R_{Q1}, the channel resistance of transistor Q1, very high with respect to R_{Q2}, the channel resistance of Q2.

The opposite situation is seen in **Figure 2-6b**. In this condition the input is HIGH and the output is LOW. Thus, Q1 is turns on and Q2 is turned off, which reverses the relationship between R_{Q1} and R_{Q2}.

The two conditions demonstrate a unique attribute of CMOS devices. Regardless of whether the output is HIGH or LOW, there is always one high resistance in series with the DC power supply (V_{CC}-V_{DD}). As a result, the normal resting current is very, very low, regardless of whether the output is HIGH or LOW. This current is typically a few microamperes (µA). The only time the current rises appreciably is during the very short time between HIGH-to-LOW or LOW-to-HIGH transitions.

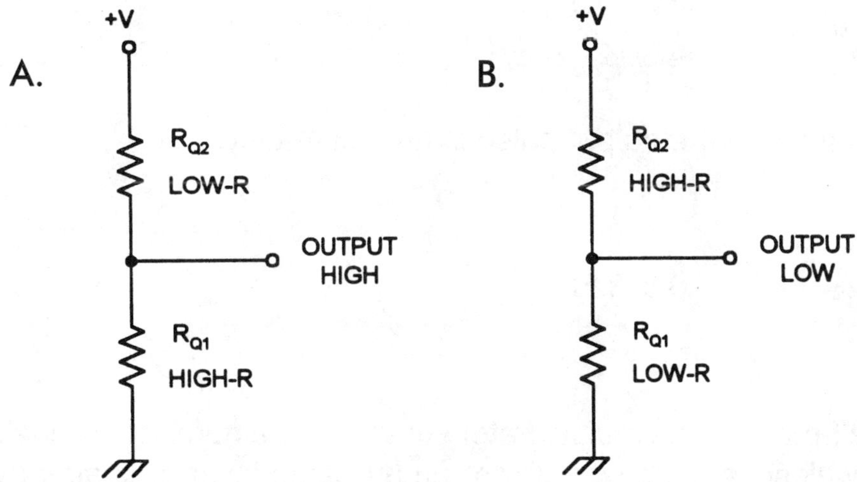

Figure 2-6. Equivalent CMOS output circuit: a) output HIGH; b) output LOW.

CMOS Half-Monostable Multivibrators

Figure 2-7 shows one of several CMOS configurations that are called half-monostables. These circuits are based on either a CMOS inverter (U1), or a CMOS buffer. The CMOS buffer is a noninverting device that provides isolation and increased drive level for the input signal (a HIGH input produces a HIGH output on a buffer). In **Figure 2-7**, a positive-going trigger produces a negative-going output pulse. If the inverter is replaced by a buffer, then a positive-going input trigger pulse produces a positive-going output pulse.

If the resistor (R1) in is tied to V+ rather than ground, then the trigger/output relationship reverses: a negative-going trigger to an inverter-based circuit produces a positive-going output pulse. Similarly, if a buffer is used for U1 a negative-going trigger pulse produces a negative-going output pulse.

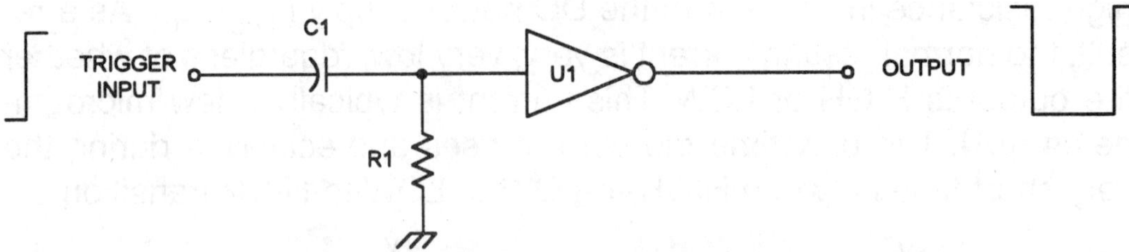

Figure 2-7. Half-monostable circuit.

The duration of the output pulse is approximately:

$$T = 0.8\ R1\ C1 \qquad \text{eq. (2-2)}$$

The half-monostable multivibrator suffers from a number of constraints, but is well-suited to cases where an available inverter is already on a printed circuit board. For example, the input trigger pulse must remain in the trigger level state for a duration longer than the duration of **Equation 2-2**. The trigger must also be noise-free. Also, the circuit cannot be retriggered until the capacitor has discharged again, so there is a long refractory period that is set by the product R1C1.

Regular CMOS Monostable Multivibrators

Figure 2-8a shows a monostable multivibrator based on the 4013B Type-D flip-flop; the timing diagram is shown in **Figure 2-8b**. A Type-D flip-flop transfers the logic level (HIGH or LOW) applied to the D-input to the Q-output when the clock (CLK) is active (in this case HIGH); the NOT-Q output assumes the opposite state as the Q output. In the circuit of **Figure 2-8a** the D-input is tied permanently HIGH, and the triggering occurs when the CLK input transitions from LOW-to-HIGH.

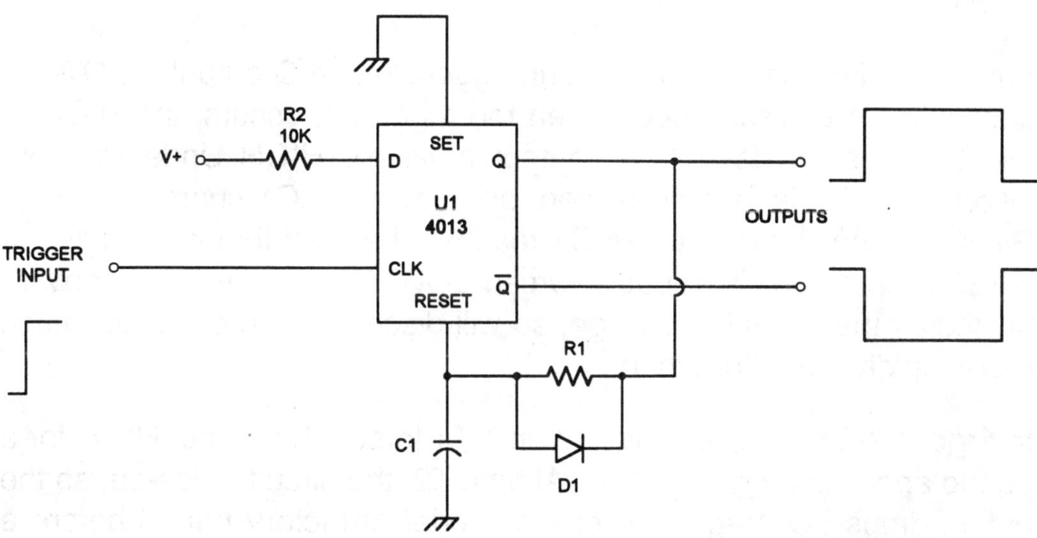

Figure 2-8a. A Type-D flip-flop used as an MMV.

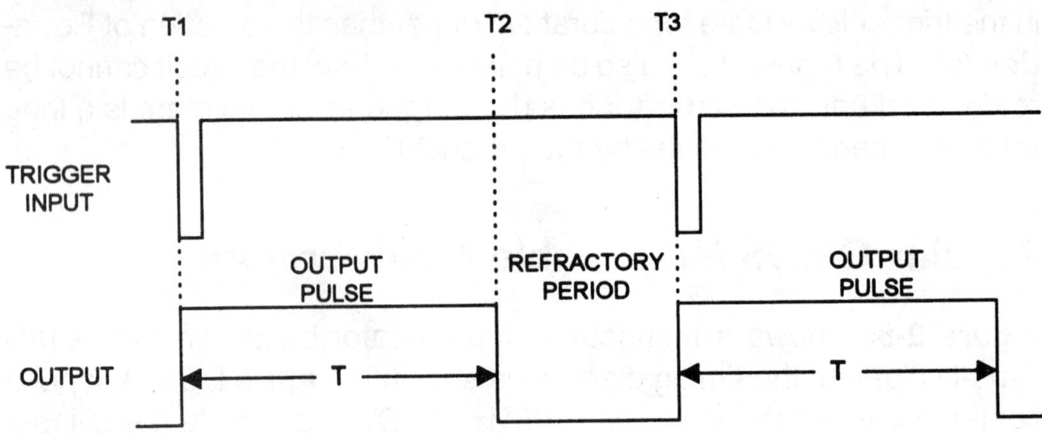

Figure 2-8b. The timing diagram.

The 4013 device used in this circuit is a direct-set type. A HIGH on the SET input forces the Q to HIGH and NOT-Q to LOW; a LOW on the SET input allows the device to operate normally. The RESET input works just the opposite: a HIGH on RESET forces Q to LOW and NOT-Q to HIGH.

When the circuit is dormant (i.e. untriggered), the Q output is LOW, so capacitor C1 is discharged. When the triggering occurs, the HIGH at the D-input is transfered to the output, making Q HIGH. Under this condition diode D1 is reverse biased, and capacitor C1 charges through R1. When the voltage across C1 reaches the level that will toggle the RESET input, the Q output is forced LOW. Now diode D1 is forward biased by the capacitor voltage, so will discharge the capacitor much more rapidly than it charged.

In **Figure 2-8b** a trigger pulse at time T1 forces the output HIGH for a period approximately 0.8R1C1. At time T2, the circuit times-out so the output drops LOW again. There is a brief refractory period before a subsequent trigger pulse can be received. The length of the refractory period is proportional to the value of C1 and the on-resistance of D1.

The 4528 dual retriggerable CMOS monostable multivibrator chip is shown in **Figure 2-9**. Pins 1 through 7 form one monostable multivibrator (MMV), while pins 9 through 15 form the other. Pin 16 is the V+ power supply, and pin 8 is the signal and DC power supply ground. Pins 1 and 15 are also grounded, and form part of the timing network.

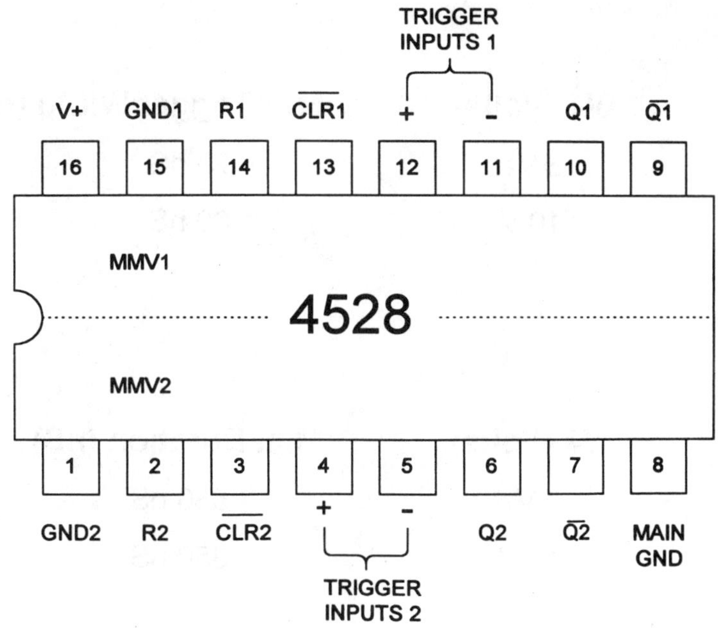

Figure 2-9. Type 4528 dual CMOS MMV package and pin-outs.

The duration of the output pulse is found from

- $T = RC$ *eq. (2-3)*

Where:

T is the pulse duration in seconds (S)

R is either R1 or R2 in ohms (Ω)

C is either C1 or C2 in farads (F)

The value of *R* can be 10 kohms to 10 megohms, and *C* should be greater than 20 pF. The minimum trigger width is found from **Table 2-1**, while the minimum duration (output pulse) is found in **Table 2-2**.

Table 2-1

V+ (Volts)	Min. Trigger Width (nS)
5 V	70 nS
10 V	30 nS

Table 2-2

V+ (Volts)	Min. Duration (nS)
5 V	550 nS
10 V	350 nS

The triggering mode is determined by the "+" and "-" Trigger Inputs. Normally, the NOT-CLR1 and NOT-CLR2 are kept HIGH. If a CLRx input is brought LOW, that forces the Q HIGH/NOT-Q LOW condition. The triggering rules are:

1. Minus (-) input HIGH: the positive (+) input is the trigger terminal. Bringing it from LOW-to-HIGH causes triggering (**Figure 2-10a**).

2. Positive (+) input LOW: the minus (-) input is the trigger, and will cause the output pulse to occur when the input makes the HIGH-to-LOW excursion (**Figure 2-10b**).

Now that we have looked at TTL and CMOS circuits, let's look at a true timer IC that works in both monostable and astable modes, the LM-555.

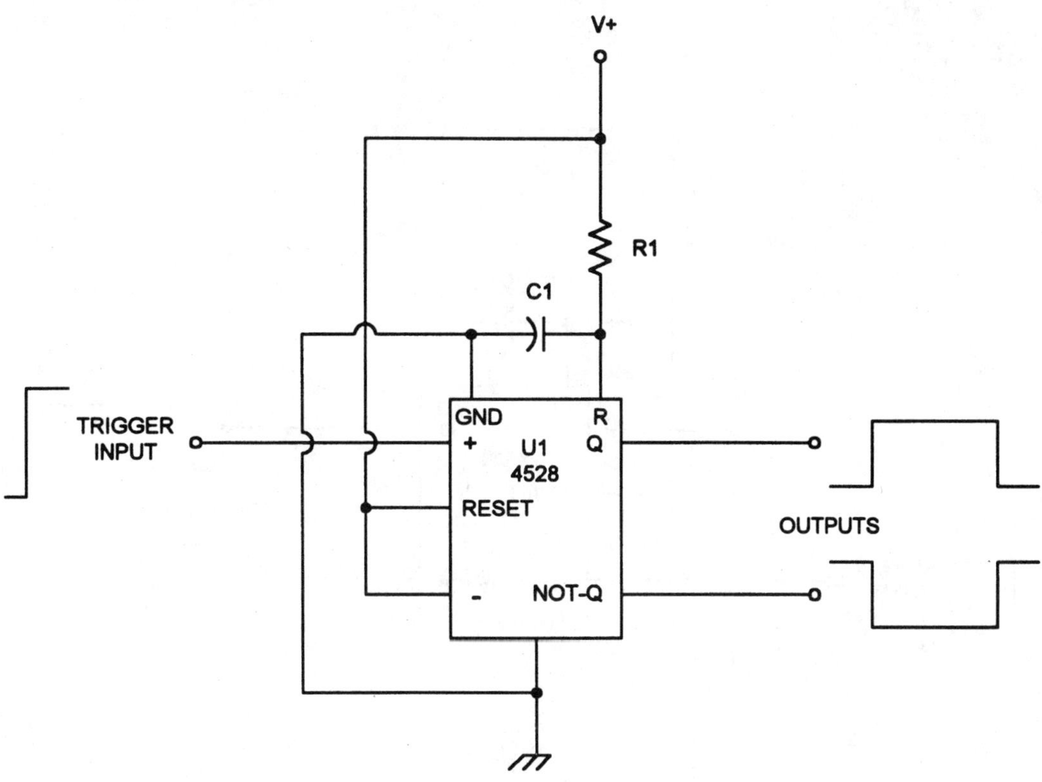

Figure 2-10a. Circuits for using the 4528 device for LOW-to-HIGH triggering.

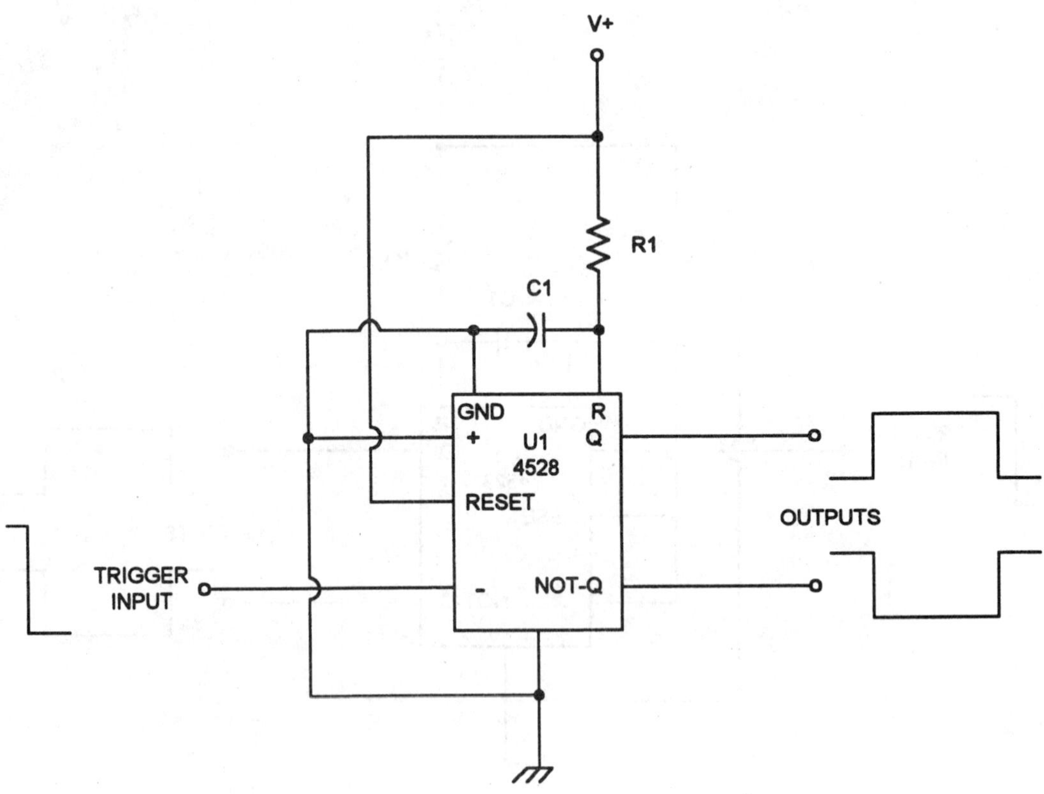

Figure 2-10b. Circuits for using the 4528 device for HIGH-to-LOW triggering.

Chapter 3

The LM-555 RC Timer IC Device

The LM-555, also known simply as the "555," has been with us for a long time, especially as electronic parts go. One of the reasons why it's been around so long is its utter usefulness. It can be operated in monostable or astable modes, and with the addition of a few external components can be made into a retriggerable monostable multivibrator. Because of the way the timer IC works, it is all but immune from pulse duration or astable frequency variations due to fluctuations in DC power supply voltage (something that many circuits cannot boast). It is easily triggered, but is not overly sensitive, and can produce an active-HIGH output that can sink or source 200 mA of current.

The original 555 device is still very much with us, but other devices were added to the line a long time ago. For example, the LM-556 device is a dual 555 in the same package. Similarly, the LMC-555 (plus similar designations by other companies) is a low-power version of the LM-555 based on CMOS logic.

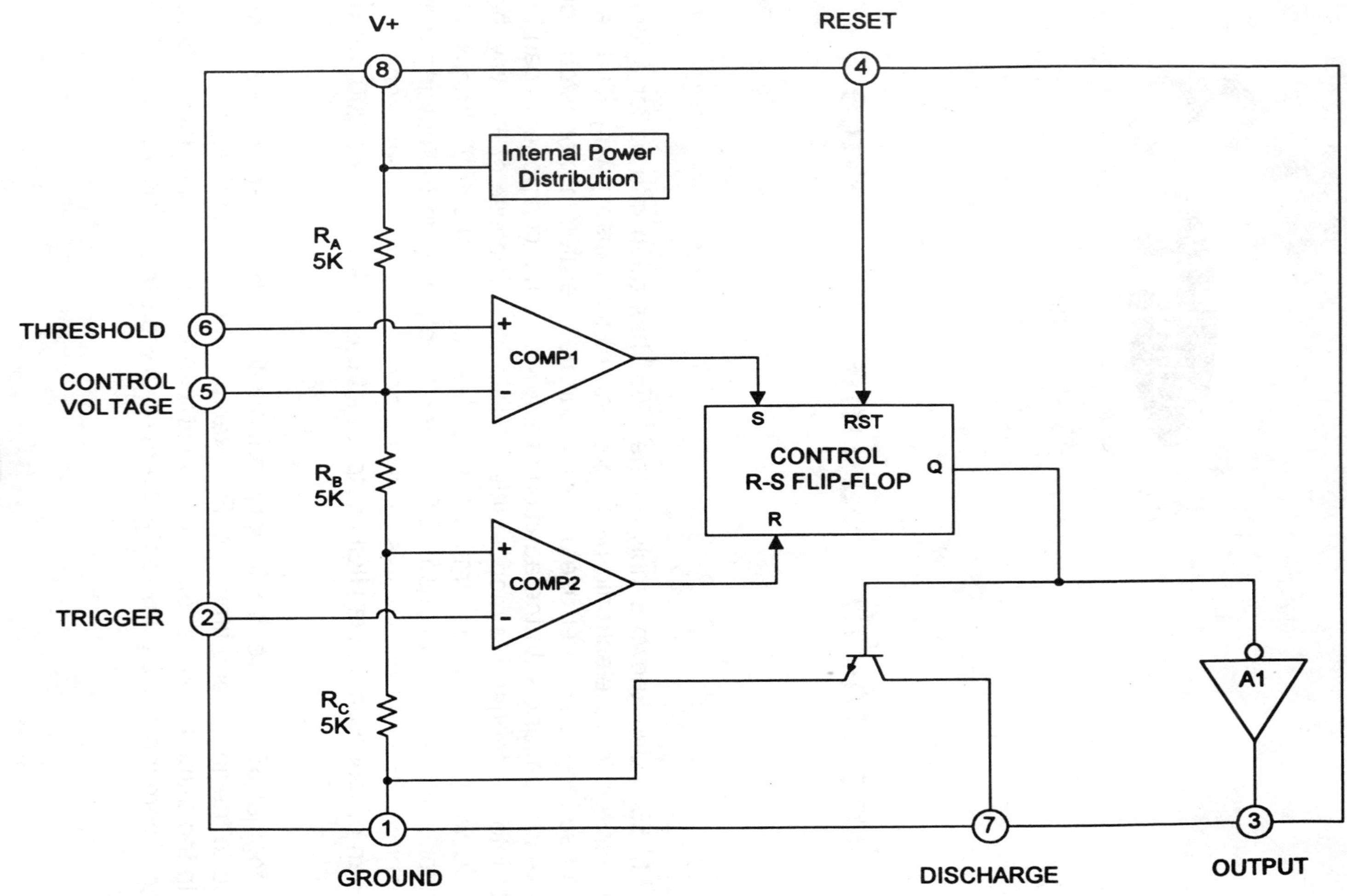

Figure 3-1a. Internal block diagram of the LM-555 timer.

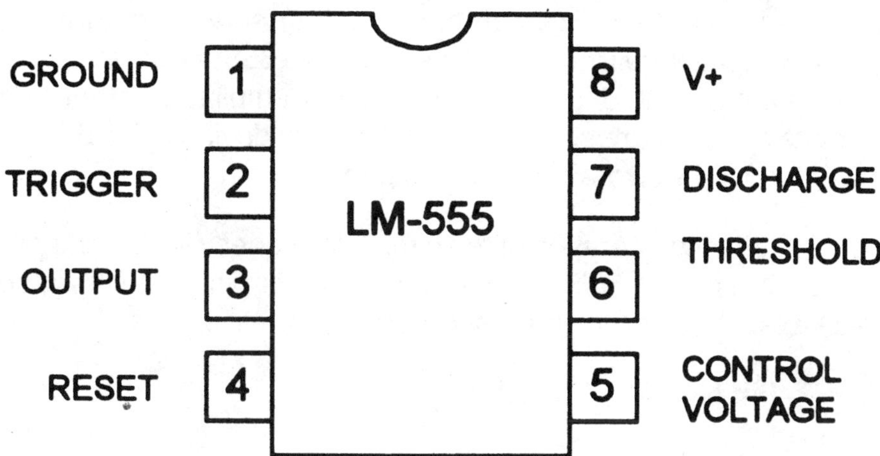

Figure 3-1b. The LM-555 package.

The LM-555 IC timer is shown in block diagram form in **Figure 3-1a**, and its package pin-outs are shown in **Figure 3-1b**. The package is an 8-pin mini-DIP (dual in-line package). This package is valid for all varieties of the 555, but not the 556. The pins shown as circles in the block diagram correspond to the pins on the IC package in the pin-out diagram.

The 555 timer represents a class of chips that are extraordinarily well-behaved, which means they are easy to apply in circuits of your own and do what is expected of them. The main reason these timers are so well-behaved is because they are based on the properties of the *series R-C timing network* and the op-amp *voltage comparator.*

The original Signetics (inventor of the 555) products included the SE-555, which operated a temperature range of -55 to +125°C, and the NE-555, which operated over the range 0 to +70°C. Several different designations are now commonly used for the 555 made by other makers, including simply 555 and LM-555, or some variant numbering.

The 555 is a multi-purpose chip that will operate at DC power supply potentials from +5 VDC to +18 VDC. The temperature stability of these devices is on the order of 50 PPM/°C (i.e., 0.005%/°C), making it a good performer over temperature excursions.

The output terminal of the 555 can either sink or source up to 200 mA of current. It is compatible with TTL devices when the 555 is operated from a +5 Vdc power supply, CMOS devices, operational amplifiers, other linear IC devices, transistors and most classes of solid-state devices. The 555 will also operate with most passive electronic components.

Before getting into the operation of the circuit, let's briefly look at the package in **Figure 3-1b**. The LM-555 is housed in an eight-pin mini-DIP integrated circuit package. The pin-outs are as follows:

Pin No.	Function
1	Ground
2	Trigger (causes output pulse when a triggering pulse is received)
3	Output (ground LOW, HIGH determined by V+)
4	Reset
5	Control voltage
6	Threshold
7	Discharge
8	V+ (positive DC voltage from +4.5 to +18 VDC)

Now let's return to **Figure 3-1a** and see what these pins actually do, and how it affects your design activities.

LM-555 Circuitry

The following stages are found in the 555: two voltage comparators (COMP1 and COMP2), a reset-set (RS) control flip-flop (which can be reset from outside the chip through pin no. 4), an inverting output amplifier (A1) and a discharge transistor (Q1). The bias levels of the two comparators are determined by a resistor voltage divider (R_a, R_b and R_c) between V+ and ground. The inverting input of COMP1 is set to 2(V+)/3, and the noninverting input of COMP2 is set to (V+)/3.

A brief description of the comparator and RS flip-flop may be in order for "...those who came in late." A voltage comparator is a differential amplifier with extremely high gain, e.g., an operational amplifier with no negative feedback resistor. Because of the high gain, only a very small signal is needed to swing the output from maximum negative (or zero) to maximum positive. The following rules apply:

1) If the same voltage is applied to -IN and +IN, the differential voltage is zero, so the output is zero.

2) If the +IN terminal is more positive than the -IN terminal, then the output is at maximum positive.

3) If the +IN terminal is less positive than the -IN terminal, then the output is at maximum negative.

An implication of these rules is that the voltage comparator (COMP1 and COMP2) serves to compare two voltages (say, V1 and V2) and issue an output that is dependent on whether V1 = V2, V1 > V2 or V1 < V2.

The RS flip-flop is designed to permit direct setting the Q and NOT-Q outputs. When the flip-flop is SET, then Q = HIGH and NOT-Q = LOW; when it is RESET then the outputs switch to Q = LOW and NOT-Q = HIGH. Bringing the RST terminal low forces the RESET condition; bringing the S terminal LOW forces the SET condition; and bringing the R terminal LOW forces the RESET condition.

In the descriptions below the term HIGH implies a level >2(V+)/3, and LOW implies a grounded condition (V = 0), unless otherwise specified in the discussion. These pins serve the following functions:

Ground (Pin No. 1). This pin serves as the common reference point for all signals and voltages in the 555 circuit both internal and external to the chip.

Trigger (Pin No. 2). The trigger pin is normally held at a potential >2(V+)/3. In this state the 555 output (pin no. 3) is LOW. If the trigger pin is brought LOW to a potential <(V+)/3, then the output (pin no. 3) abruptly switches to the HIGH state. The output remains HIGH as long as pin no. 2 is LOW, but the output does not necessarily revert back to LOW immediately after pin no. 2 is brought HIGH again (see operation of the Threshold input below).

Output (Pin No. 3). The output pin of the 555 is capable of either sinking or sourcing current up to 200 mA. This operation is in contrast to other IC devices in which the outputs of various devices will either sink or source current, but not both. Whether the 555 output operates as a sink or a source depends on the configuration of the external load. **Figure 3-2** shows both types of operation.

In **Figure 3-2a** the external load R_L is connected between the 555 output and V+. Current only flows in the load when pin no. 3 is LOW. In that condition the external load is grounded through pin no. 1 and a small internal source resistance, R_{s2}. In this configuration the 555 output is a current sink.

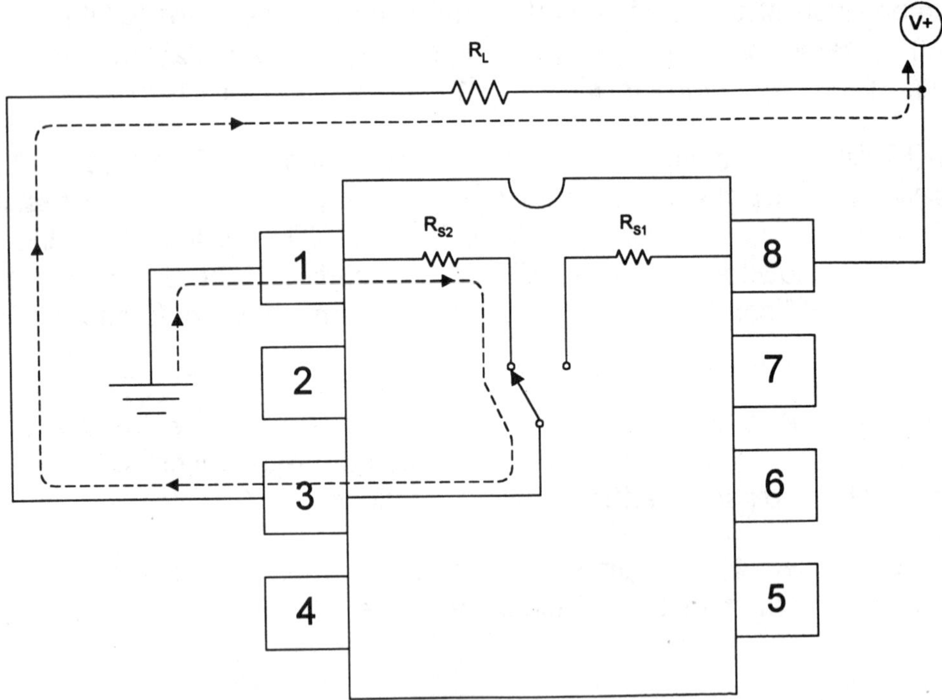

Figure 3-2a. Output LOW condition.

The operation depicted in **Figure 3-2b** is for the case where the load is connected between pin no. 3 of the 555 and ground. When the output is LOW the load current is zero. But when the output is HIGH, however, the load is connected to V+ through a small internal resistance R_{s1} and pin no. 8. In this configuration the output serves as a current source.

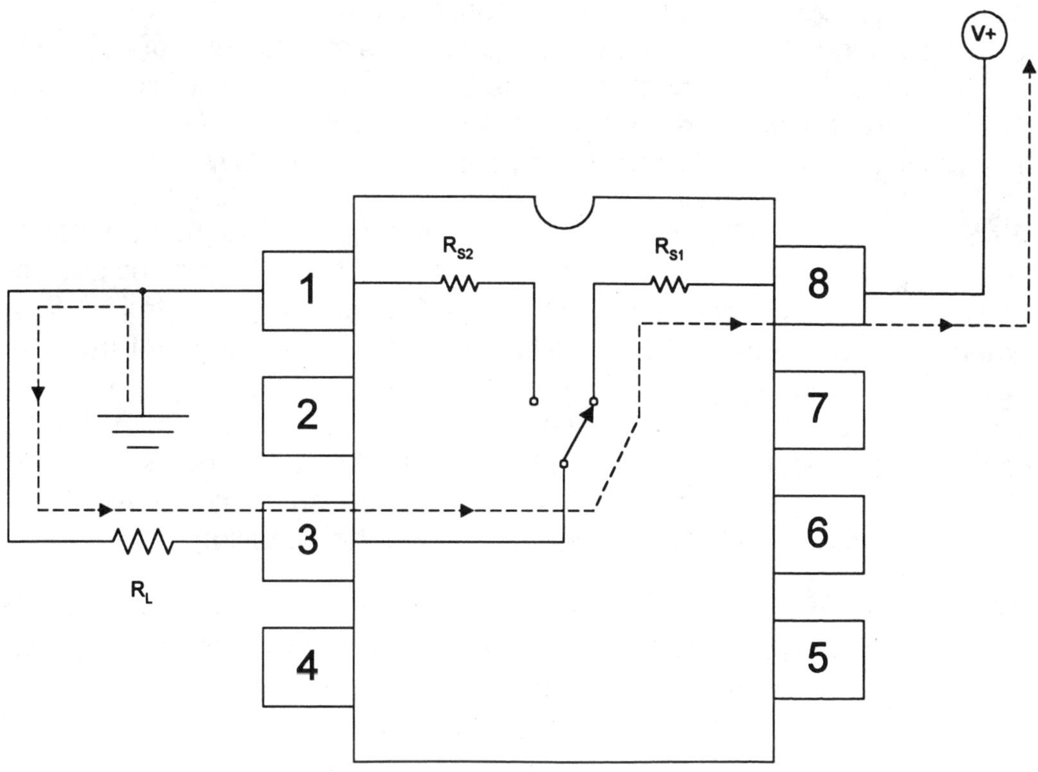

Figure 3-2b. Output HIGH condition.

Reset (Pin No. 4). The reset pin is connected to a preset input of the 555 internal control flip-flop. When a LOW is applied to pin no. 4 the output of the 555 (pin no. 3) switches immediately to a LOW state. In normal operation it is common practice to connect pin no. 4 to V+ in order to prevent false resets from noise impulses. This connection forces the RST on the RS flip-flop permanently inactive.

Control Voltage (Pin No. 5). This pin normally rests at a potential of 2(V+)/3 due to an internal resistive voltage divider (see R_A through R_C in **Figure 3-1a**). Applying an external voltage to this pin, or connecting a resistor to ground, will change the duty cycle of the output signal. If not used, then pin no. 5 should be decoupled to ground through a 0.01 μF to 0.1 μF capacitor.

Threshold (Pin No. 6). This pin is connected to the noninverting input (+IN) of comparator COMP1, and is used to monitor the voltage across the capacitor in the external RC timing network. If pin no. 6 is at a potential of <2(V+)/3, then the output of the control flip-flop is LOW, and the output (pin no. 3) is HIGH. Alternatively, when the voltage on pin no. 6 is > 2(V+)/3, then the output of COMP1 is HIGH and chip output (pin no. 3) is LOW.

Discharge (Pin No. 7). The discharge pin is connected to the collector of NPN transistor Q1, and the emitter of Q1 is connected to the ground pin (no. 1). The base of Q1 is connected to the NOT-Q output of the control flip-flop. When the 555 output is HIGH, the NOT-Q output of the control flip-flop is LOW, so Q1 is turned off. The c-e resistance of Q1 is very high under this condition, so does not appreciably affect the external circuitry. But when the control flip-flop NOT-Q output is HIGH, however, the 555 output is LOW and Q1 is biased hard on. The c-e path is in saturation, so the c-e resistance is very low. Pin no. 7 is effectively grounded under this condition

V+ Power Supply (Pin No. 8). The DC power supply is connected between ground (pin no. 1) and pin no. 8, with pin no. 8 being positive. In good practice a 0.1 μF to 10 μF decoupling capacitor will normally be used between pin no. 8 and ground.

Once the basic functioning (including internal workings) of the LM-555, and the action of each of the pins is understood, then it is possible to use the LM-555 device in designs of your own. If you are unclear about the operation of the device, then please reread the above material until you know it well. In the chapters to follow we will take a look at the monostable and astable configurations of the LM-555, as well as some of their respective applications.

Chapter 4

555 Monostable Operation

Recall from Chapter 1 that a one-shot circuit, also called the monostable multivibrator (MMV), produces a single output pulse for each input trigger pulse. **Figure 4-1** shows one-shot operation of the LM-555 IC timer device. The output remains dormant in the LOW condition (at or very near 0V) until a trigger pulse is received. At that time, the output snaps HIGH (V_o is just a little below the chip V+) and remains in that state for a specified time duration *T*. When *T* expires, the output drops back to LOW and remains there until another trigger pulse is received. The output pulse duration must be greater than the trigger pulse duration ($T > t_p$).

The LM-555 is a negative-going trigger device. This means that the trigger input is kept at a high potential near V+. A successful trigger pulse must drop from V+ to less than (V+)/3, but higher than 0V (negative excursions may ruin the chip, although this is not a certainty). This action is shown in the bottom waveform in **Figure 4-1**.

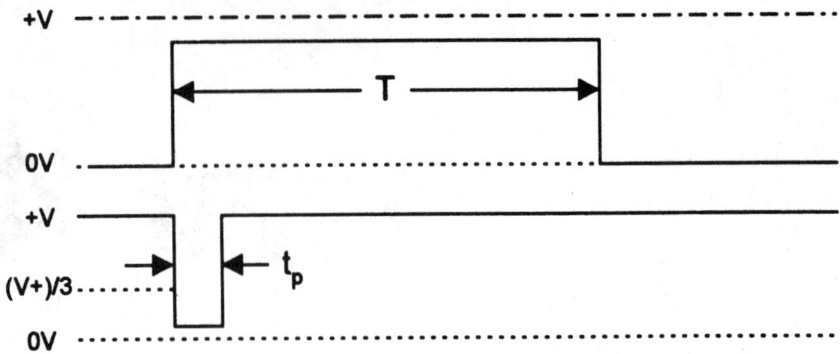

Figure 4-1. Timing waveforms of the LM-555 monostable multivibrator.

Figure 4-2a shows the circuit for the one-shot circuit, while its timing waveforms are shown in **Figure 4-2b**. The timing waveforms show the time relationship between the input trigger pulse, the timing capacitor voltage and the output pulse.

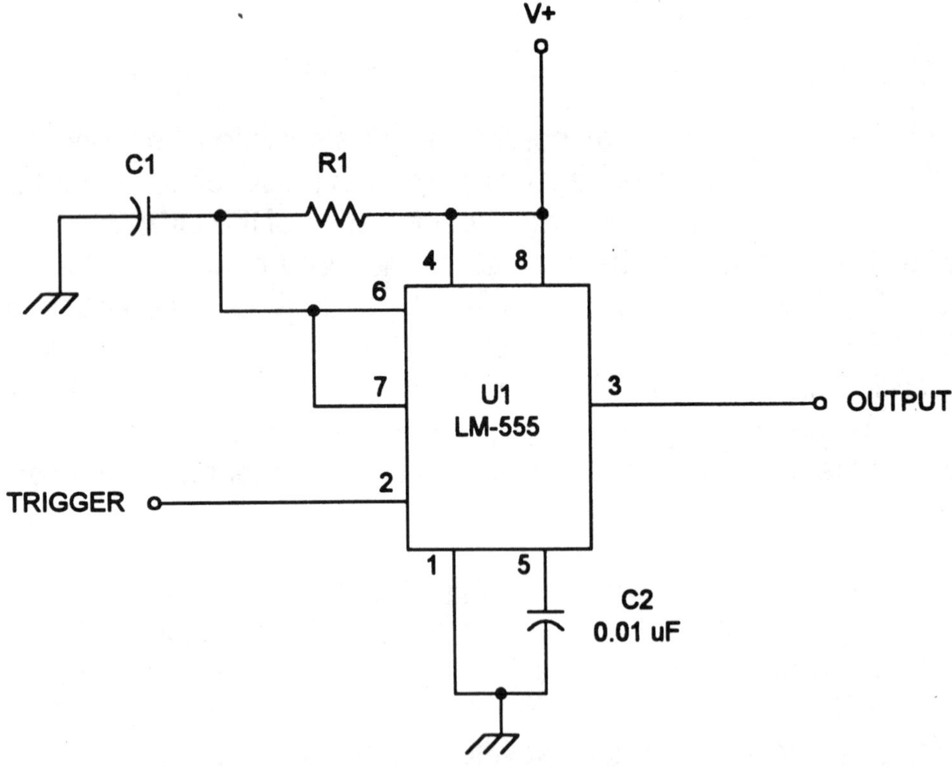

Figure 4-2a. LM-555 monostable multivibrator circuit.

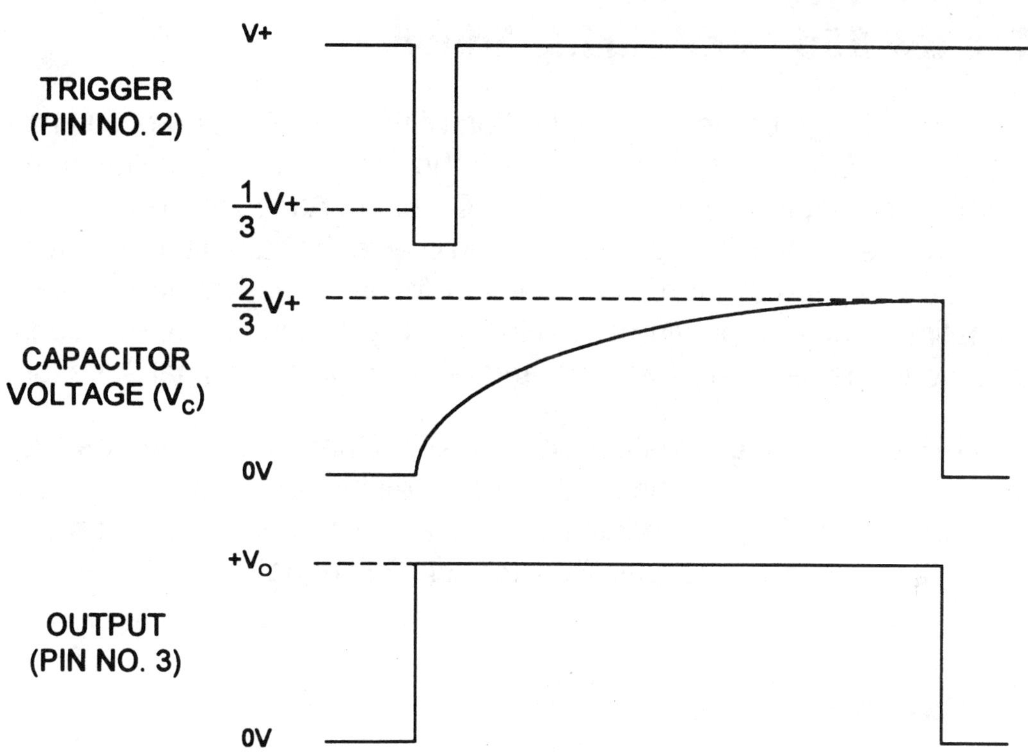

Figure 4-2b. Circuit timing waveforms showing capacitor charge voltage.

The control voltage (pin no. 5) is bypassed to ground by a 0.01 μF capacitor, although anything from that value to 0.1 μF will do nicely. The timing network consists of R1 and C1, and is connected so that the capacitor voltage V_c appears on the threshold terminal of the 555. Note also that the Discharge pin (no. 7) is also connected across the capacitor.

This pin is connected to an internal transistor collector (see Chapter 3 for details of the LM-555 innards) that will go LOW when the RS flip-flop goes to the condition where NOT-Q goes HIGH. At that time the DC path through the transistor discharges the capacitor, starting the timing sequence over again. The discharge of C1 occurs when the capacitor voltage reaches 2(V+)/3.

The LM-555 MMV Timing Equation

From knowledge of the MMV operation of the LM-555, and of RC timing circuits (Chapter 1), we can derive the equation for calculating the duration *T* from the values of R1 and C1 used. Although I realize that many readers don't like derivations, and one technical editor maintains that we lose about half our readership for each derivation (lie! lie!), sometimes they do provide some insight. If you don't like derivations, then go to the end of this section and take a look at the final equation.

The basic RC timing equation discussed in Chapter 1 is the basis for the LM-555 timing equation, and relates the time required for a capacitor voltage in an RC network to rise from a stating point (V_{C1}) to an end point (V_{C2}), given a particular RC product. This equation is:

- $$T = -R\,C\,Ln\left[\frac{V - V_{C2}}{V - V_{C1}}\right]$$ *eq. (4-1)*

In the LM-555 timer IC the source voltage is the chip V+, the starting voltage is zero (i.e. the discharged value of the capacitor voltage), and the trip-point for the internal comparator (COMP1) is 2/3-(V+). **Equation 4-1** can therefore be rewritten to take into account these values:

- $$T = -R1\,C1\,Ln\left[\frac{(V+) - (2(V+)/3)}{(V+) - 0}\right]$$ *eq. (4-2)*

- $$T = -R1C1\ Ln(1 - 0.667)$$ *eq. (4-3)*

- $T = -R1C1\ Ln(0.333) = 1.1R1C1$ *eq. (4-4)*

Thus, the timing equation for the LM-555 device in the monostable multivibrator mode is:

- $T = 1.1\,R1\,C1$ *eq. (4-5)*

Where:

T is the output pulse duration in seconds

R1 is in ohms (Ω)

C1 is in farads (F)

Note: If you use microfarads (μF) for C1, then T is expressed in microseconds (μS).

Because it is usually easier to obtain standard value resistors in a wide range of values, and variable resistors are both available and cheap, the general procedure is to select a trial value for C1 from the standard table, and then find the value of resistance that is required to generate the desired time. For example, to make a one millisecond (1 ms = 0.001 s = 1,000 μs) one-shot, select 0.01 μF and solve **Equation 4-1** for R1 = T/1.1C1 = 1000 μS/(1.1 $\times$ 0.01 μF) = 90,909 ohms. This value is close enough to 91k (a standard value) to make it a reasonable choice.

Input Triggering Methods

The 555 MMV circuit triggers by bringing pin no. 2 from a positive voltage down to a level <(V+)/3. Triggering can be accomplished by applying a pulse from an external signal source, or through other means. **Figure 4-3** shows the circuit for a simple pushbutton switch trigger circuit. A pull-up resistor (R2) is connected between pin no. 2 and V+. If normally-open (N.O.) pushbutton switch S1 is open, then the trigger input is held at a potential very close to V+. But when S1 is closed, pin no. 2 is brought LOW to ground potential. Because pin no. 2 is now at a potential less than (V+)/3 the 555 MMV will trigger. This circuit can be used for contact debouncing.

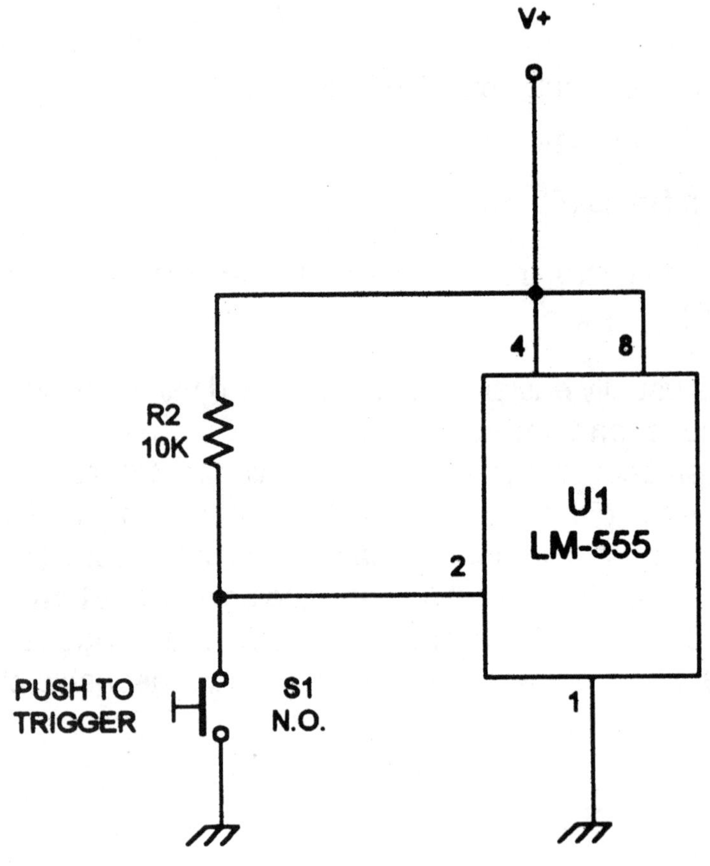

Figure 4-3. Pushbutton triggering of MMV.

A circuit for inverting the trigger pulse applied to the 555 is shown in **Figure 4-4**. In this circuit an NPN bipolar transistor is used in the common emitter mode to invert the pulse. Again, a pull-up resistor is used to keep pin no. 2 at V+ when the transistor is turned off. But when the positive polarity trigger pulse is received at the base of transistor Q1, the transistor saturates and this forces the collector (and pin no. 2 of the 555) to near ground potential.

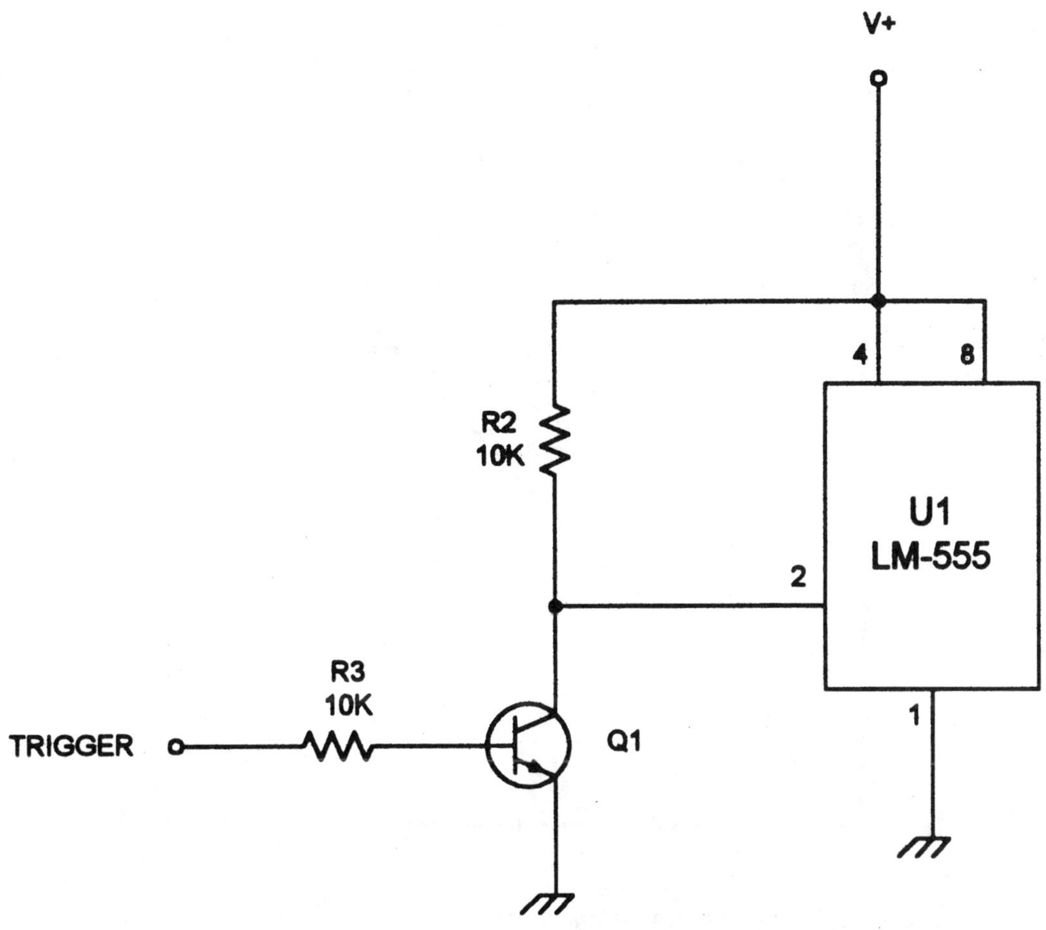

Figure 4-4. Using NPN driver to invert the trigger input, allowing triggering by positive-going pulses.

Figure 4-5a shows two AC-coupled versions of the trigger circuit. In these circuits a pull-up resistor keeps pin no. 2 normally at V+. But when a pulse is applied to the input end of capacitor C3, a differentiated version of the pulse is created at the trigger input of the 555. Diode D1 clips the positive-going spike to 0.6 - 0.7V, passing only the negative-going pulse to the 555. If the negative-going spike can counteract the positive bias provided by R2 sufficiently to force the voltage lower than (V+)/3, then the 555 will trigger. A pushbutton switch version of this same circuit is shown in **Figure 4-5b**.

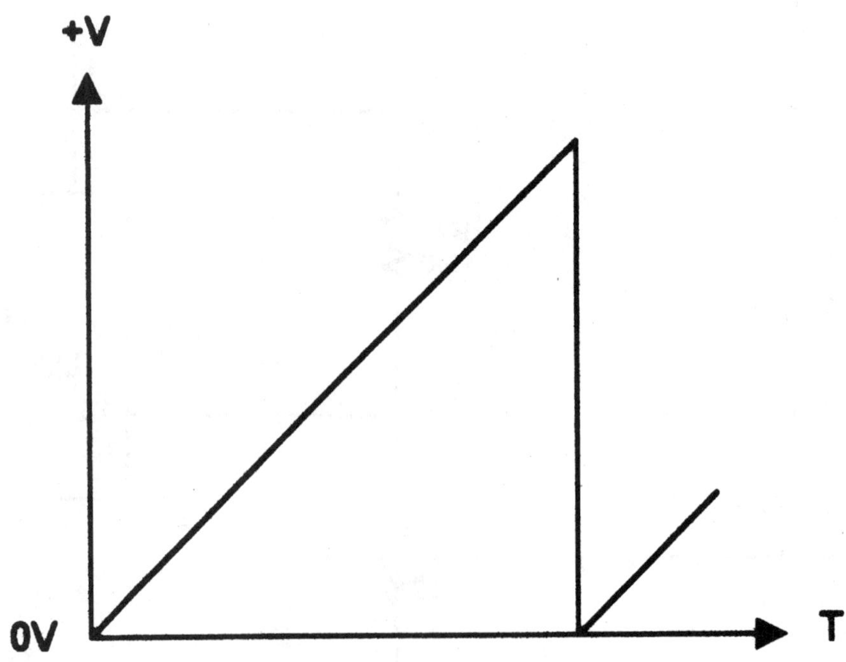

Figure 4-5a. AC-coupled triggering.

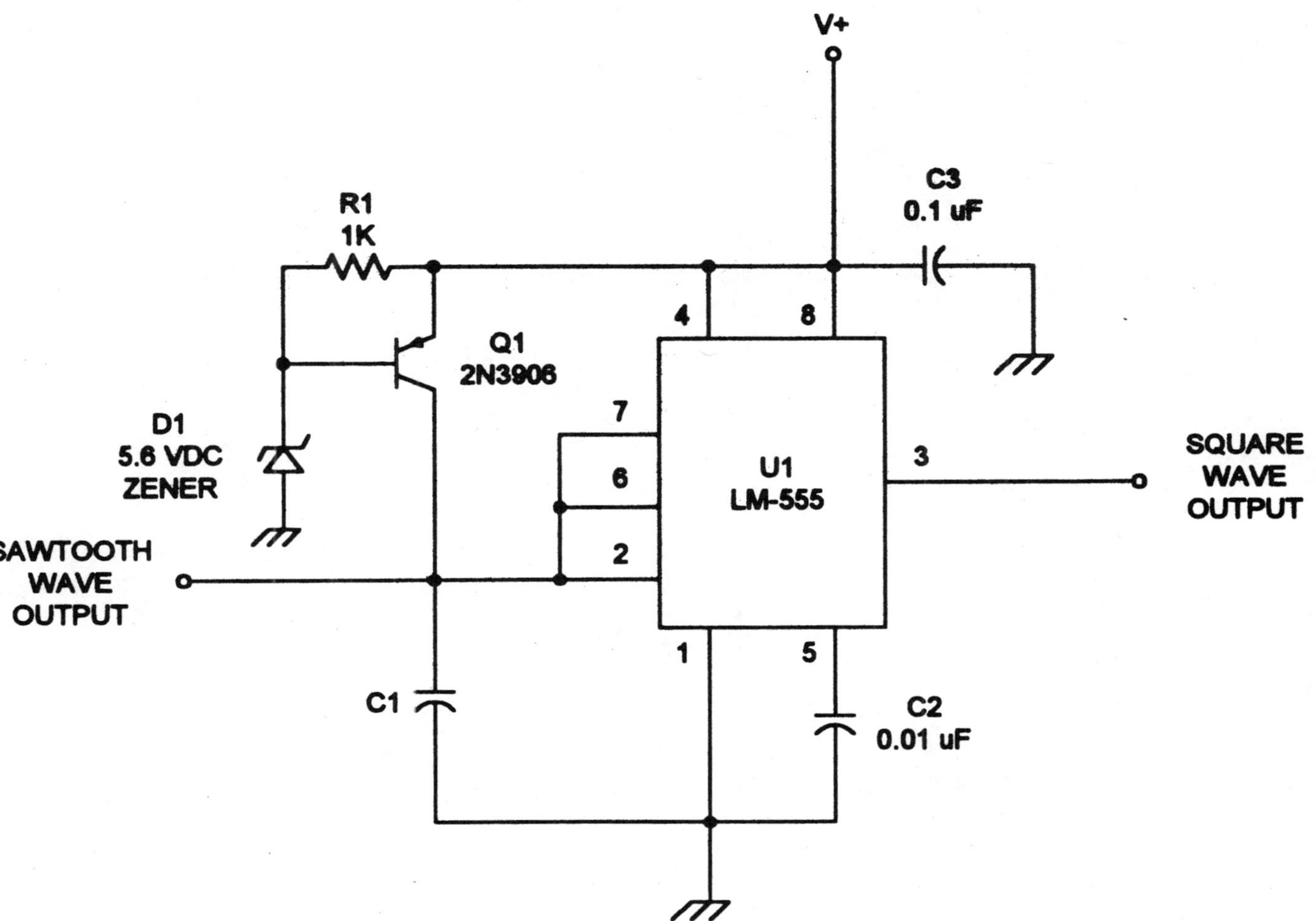

Figure 4-5b. Pushbutton version of AC-coupled triggering.

Chapter 5

555 Astable Multivibrator Circuits

An astable multivibrator is a multivibrator circuit that has two output states, but neither of them is stable (hence "astable"). The general case is shown in **Figure 5-1a**. The square wave is symmetrical above and below the baseline, and the high and low times are the same. Thus, the square oscillates between $-V_O$ and $+V_O$, where $ABS(-V_O) = ABS(+V_O)$.

A more restrictive case is shown in **Figure 5-1b**. This waveform oscillates back and forth between 0V and some output voltage V_O. In most circuits, the value of V_O is a little bit below V+, which is also true of the LM-555 device.

The LM-555 device is relatively easy to make into an astable multivibrator and is well-behaved. That is to say, it does what you tell it to do (within the rules, of course). In the next section we will look at the astable configuration of the LM-555 along with a couple of representative application circuits.

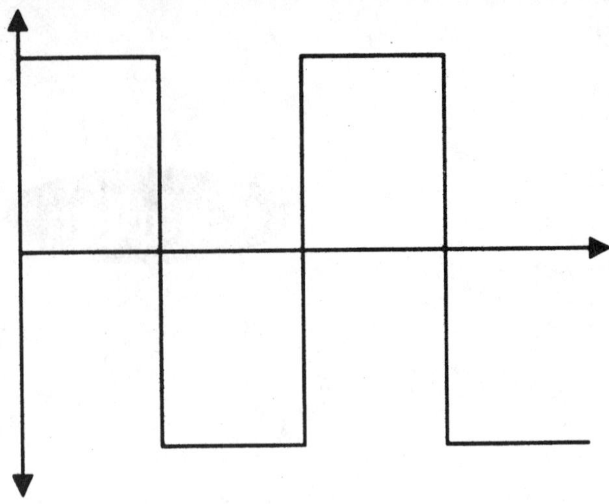

Figure 5-1a. A true square wave from an astable multivibrator.

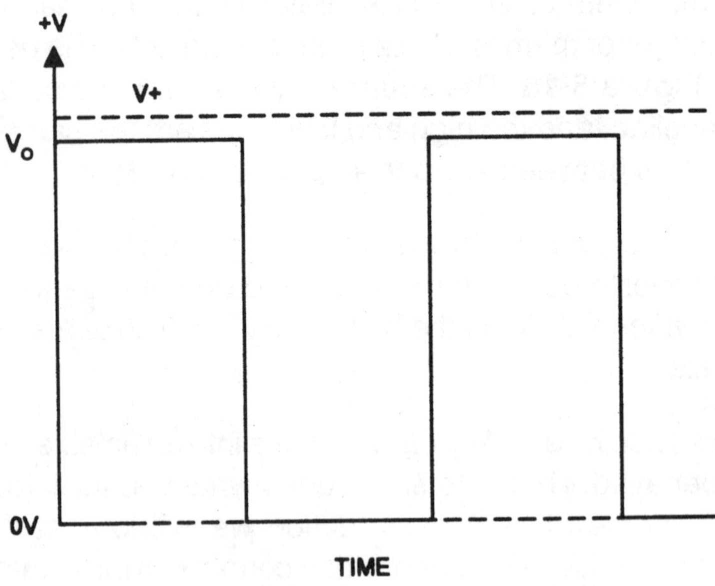

Figure 5-1b. A Unipolar square wave of the type produced by the LM-555.

Astable Operation of the 555 IC Timer

The 555 can be connected to produce a variable duty cycle AMV circuit. A version of the circuit showing the internal stages of the 555 is shown in **Figure 5-2a**, while the circuit as it normally appears in schematic drawings is shown in **Figure 5-2b**. The factor that makes this circuit an AMV is that the threshold and trigger pins (6 and 2) are connected together, forcing the circuit to be self-retriggering.

Under initial conditions at turn-on the voltage across timing capacitor C1 is zero, while the biases on COMP1 and COMP2 are (as usual) set to 2(V+)/3 and (V+)/3, respectively, by the internal resistor voltage divider (R_A, R_B and R_C). The output of the 555 is HIGH under this condition, so C1 begins to charge through the combined resistance [R1 + R2]. On discharge, however, transistor Q1 shorts the junction of R1 and R2 to ground, so the capacitor discharges through R2 only. The result is the waveform shown in **Figure 5-2c**. The time that the output is HIGH is t1; the LOW time is t2. The period *T* of the output square wave is the sum of these two durations: T = (t1 = t2).

As with all similar RC-timed circuits the equation that sets oscillating frequency is determined from **Equation 5-1**:

$$T = -R1C1 \text{ LN} \left[\frac{V - V_{C2}}{V - V_{C2}} \right] \qquad \textit{eq. (5-1)}$$

For the case where the output is HIGH (t2), the resistance R is [R1 + R2], and the capacitance *C* is C1. Because of the internal biases of the voltage comparator stages of the 555, the capacitor will charge from (V+)/3 to 2(V+)/3, and then discharge back to (V+)/3 on each cycle. Thus, **Equation 5-1** can be rewritten in the form:

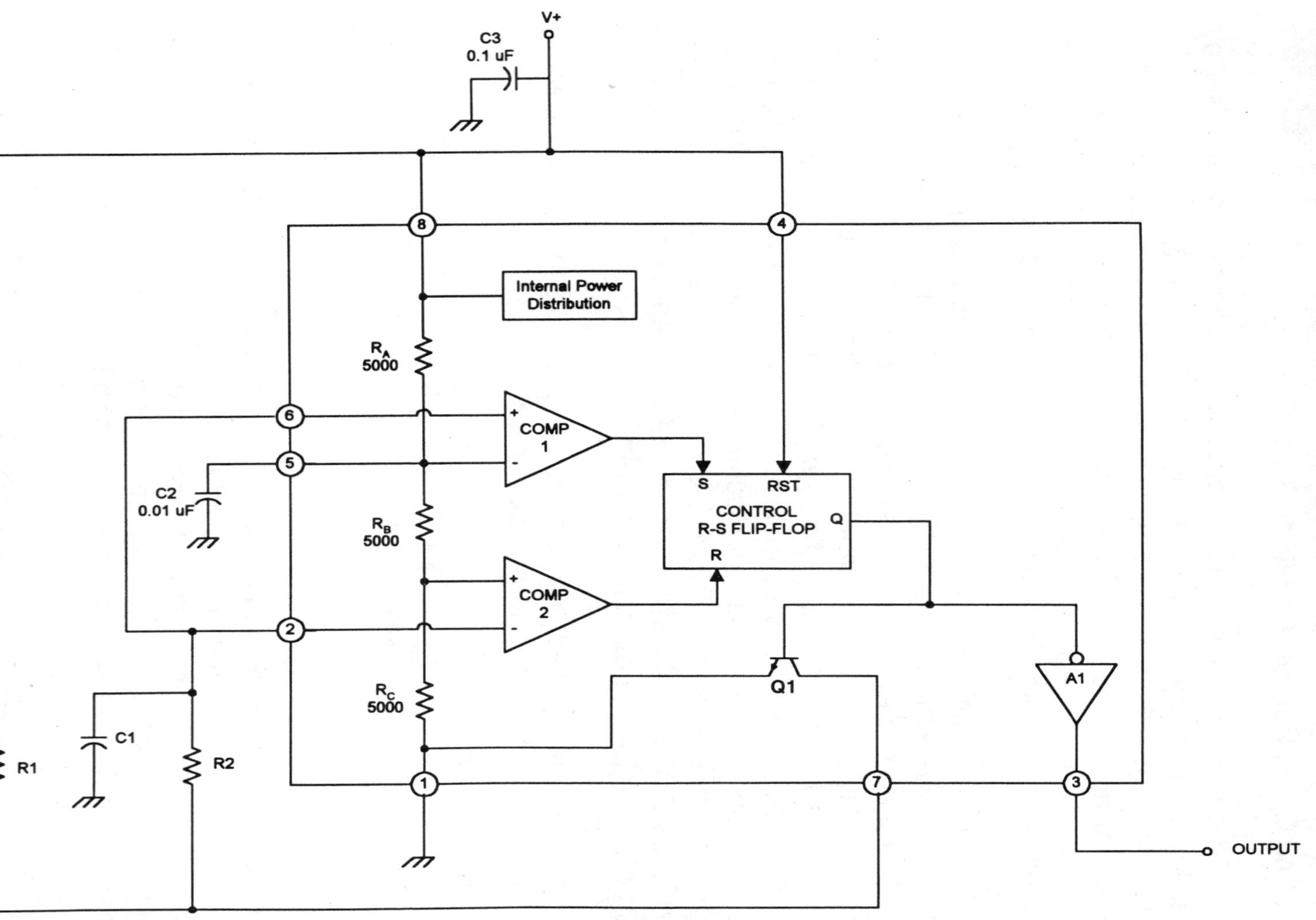

Figure 5-2a. The LM-555 astable multivibrator circuit showing internal workings of the IC.

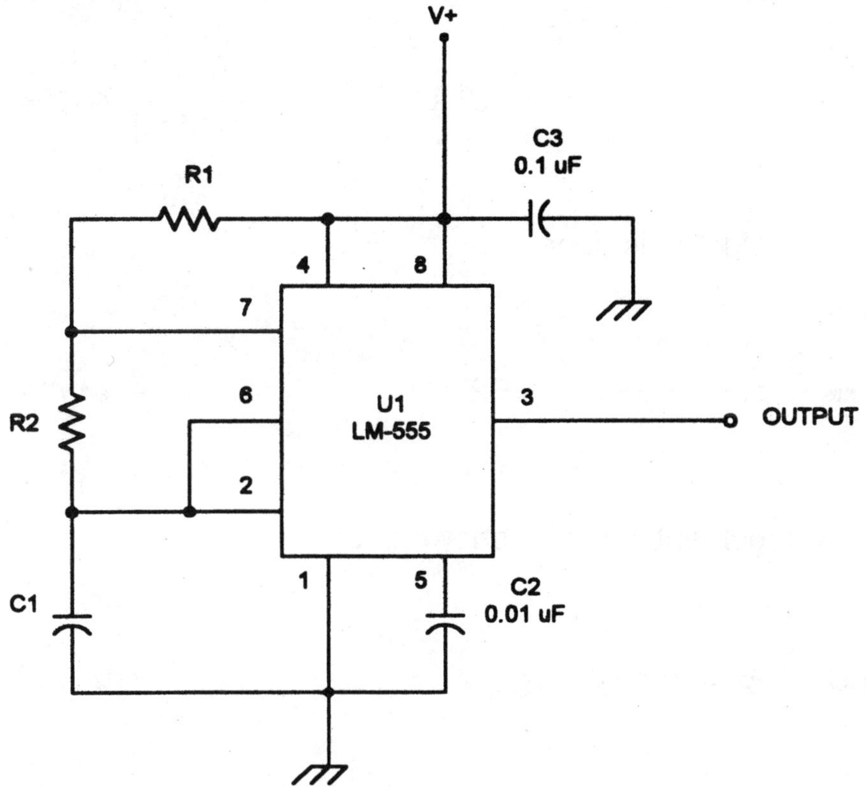

Figure 5-2b. The LM-555 astable multivibrator circuit as it appears in schematic diagrams.

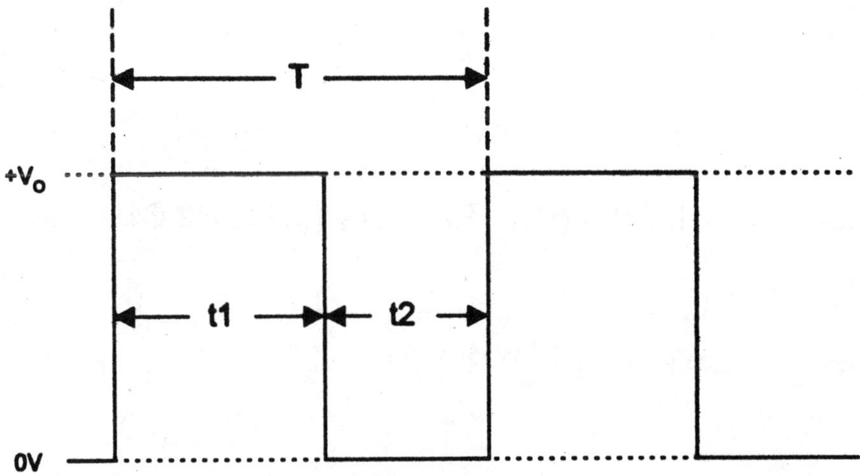

Figure 5-2c. The LM-555 astable multivibrator output waveform showing time relationships.

$$t1 = -(R1 + R2)\,C1 \;\mathrm{LN}\left[\frac{(V+) - 2(V+)/3}{(V+) - (V+)/3}\right] \qquad eq.\ (5\text{-}2)$$

or, once the algebra is done:

$$t1 = 0.695\ (R1 + R2)\ C1 \qquad eq.\ (5\text{-}3)$$

By similar argument it can be shown that:

$$t2 = 0.695\ R2\ C1 \qquad eq.\ (5\text{-}4)$$

For the total period *T*:

$$T = t1 + t2 \qquad eq.\ (5\text{-}5)$$

$$T = [(0.695\ (R1 + R2)\ C1] + [0.695\ R2\ C1] \qquad eq.\ (5\text{-}6)$$

$$T = 0.695\ (R1 + 2\ R2)\ C1 \qquad eq.\ (5\text{-}7)$$

Equation 5-7 defines the period of the output square wave. In order to find the frequency of oscillation take the reciprocal of **Equation 5-7**:

- $$F = \frac{1}{T}$$ eq. (5-8)

- $$F = \frac{1.44}{(R1 + 2R2)C1}$$ eq. (5-9)

The Duty Cycle of 555 Astable Multivibrator

Time segments t1 and t2 are not equal in most cases, so the charge and discharge times for capacitor C1 are also not equal (see **Figure 5-2d**). The duty cycle of the output signal is the ratio of the HIGH period to the total period (t1/T).

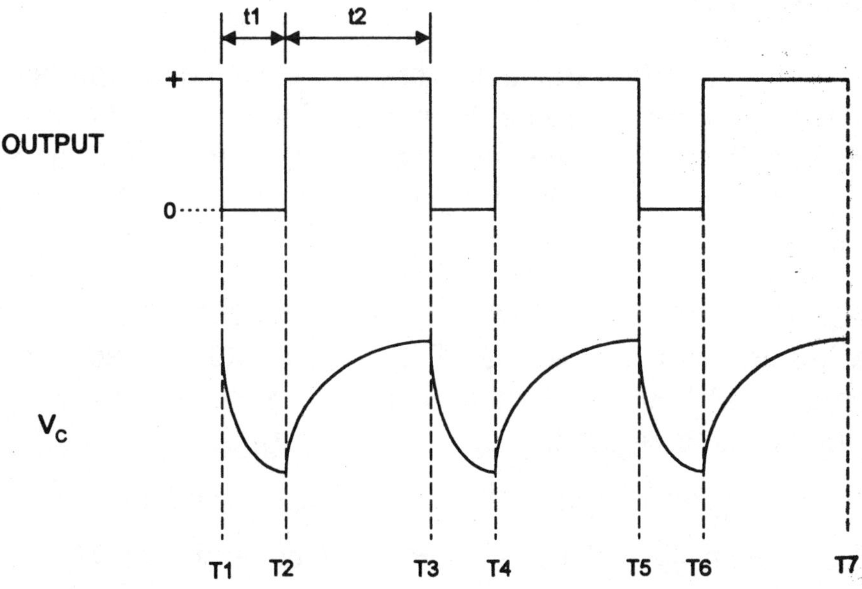

Figure 5-2d. The LM-555 astable multivibrator Timing waveforms of the LM-555 astable multivibrator.

Expressed as a percent:

$$\%DC = \frac{R1 + R2}{R1 + 2R2} \qquad \text{eq. (5-10)}$$

Various methods are used for varying the duty cycle. First, a voltage can be applied to pin no. 5 (Control Voltage). Second, a resistance can be connected from pin no. 5 to ground. Both of these tactics have the effect of altering the internal bias voltages applied to the comparator. Alternatively, one can also divide the external resistances R1 and R2 into three values. **Figure 5-3** shows a variable duty factor 555 AMV that uses a potentiometer (R3) to vary the ratio of the charge and discharge resistances.

Synchronized Operation of 555 Astable Multivibrator

A synchronized AMV operates in a manner in which its operating frequency is locked to an external frequency. The horizontal and vertical deflection oscillators in a television receiver operate in this manner because they are locked to the sync pulses transmitted by the TV broadcast station.

A method for locking in the oscillating frequency of the 555 is shown in **Figure 5-4**. In this circuit a 7400 TTL NAND gate is used to sample both the 555 AMV output signal and the input sync signal. The properties of the NAND gate are these:

1. If either input is LOW, then the output is HIGH.
2. Both inputs must be HIGH for the output to be LOW.

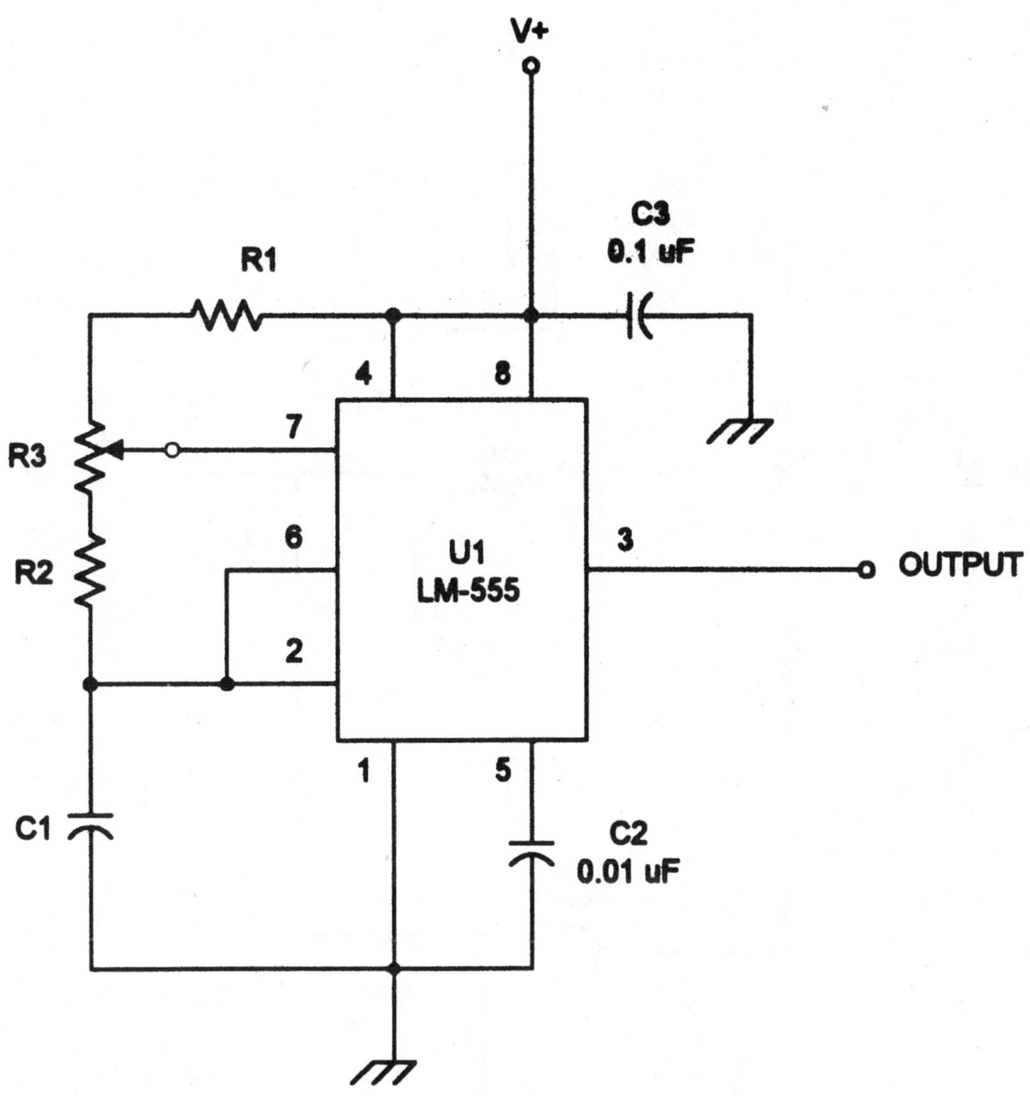

Figure 5-3. Variable duty-cycle LM-555 astable multivibrator.

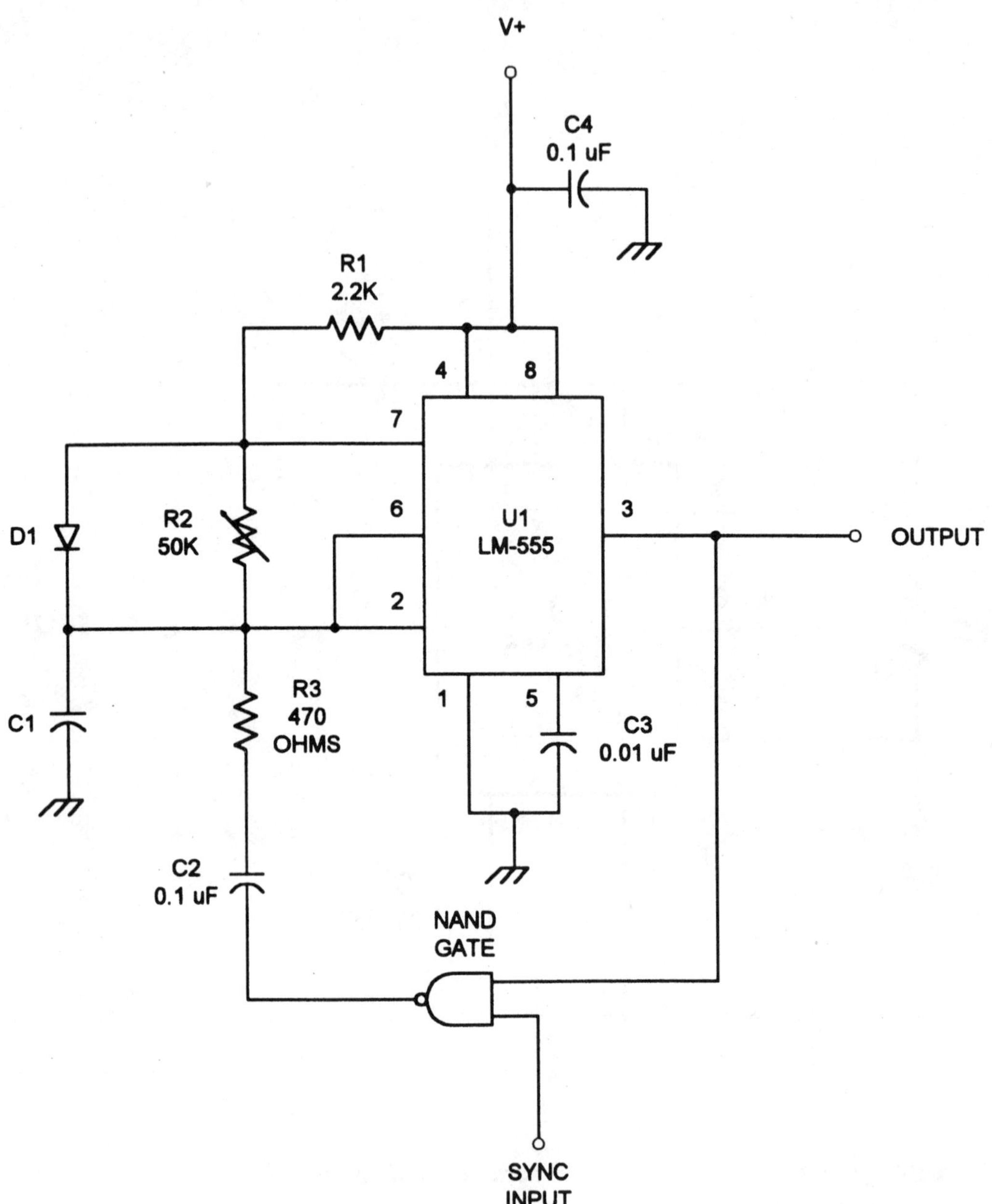

Figure 5-4. Synchronous operation of the LM-555 astable multivibrator.

Because the output of the NAND gate is applied to the timing circuit of the 555 (through waveshaping network R3/C2), it will affect the relative timing of the circuit. This circuit is analogous to a mechanical pendulum oscillator which has an external non-resonant forcing frequency applied. The circuit will lock to the new frequency if it is reasonably close to the natural oscillating frequency, or an integer harmonic of the natural frequency. Students interested in modern chaos theory might want to investigate the behavior of this and similar circuits at sync frequencies away from the natural frequency or its harmonics and subharmonics.

A 555 Sawtooth Generator Circuit

A sawtooth waveform (**Figure 5-5a**) rises linearly to some value, and then drops abruptly back to the initial conditions.

The circuit for a 555-based sawtooth generator (**Figure 5-5b**) is simple, and is based on the 555 timer IC. The basic circuit is the monostable multivibrator configuration of the 555, in which one of the timing resistors is replaced with a transistor operated as a current source (Q1). Almost any audio small signal PNP silicon replacement transistor can

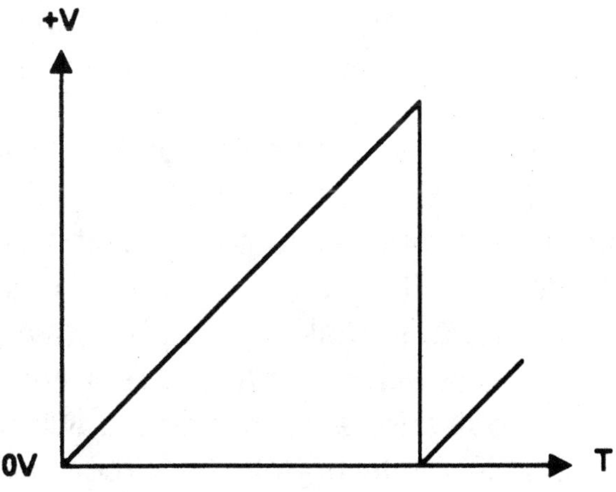

Figure 5-5a. A sawtooth waveform.

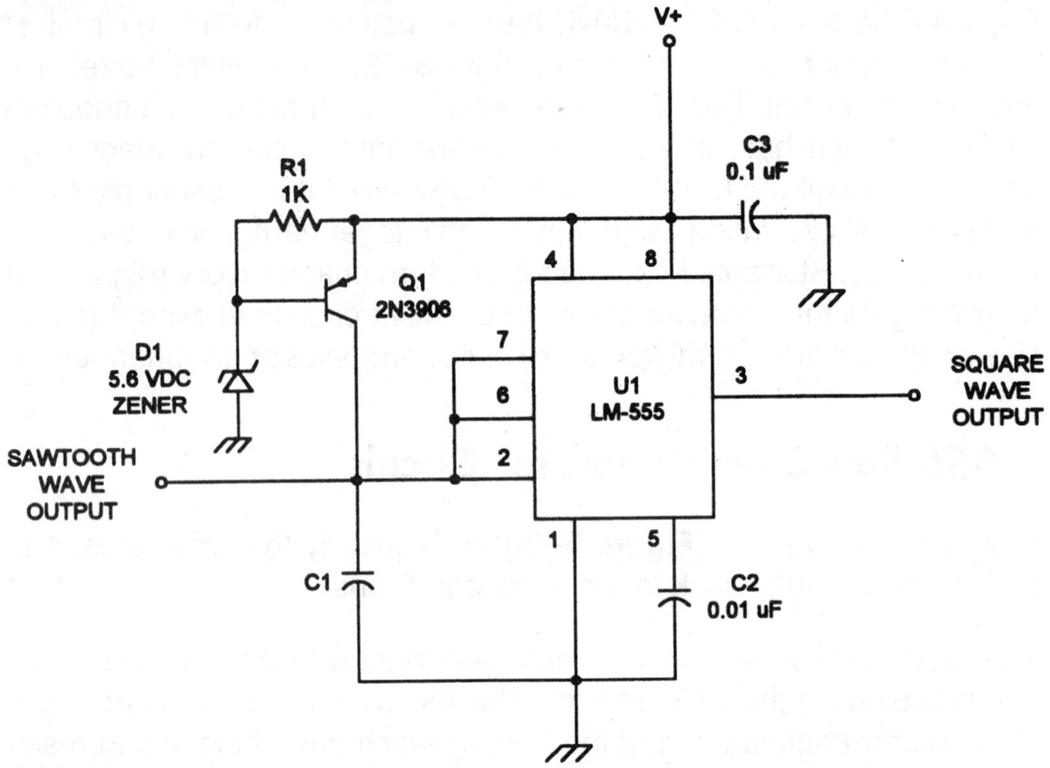

Figure 5-5b. A sawtooth generator based on the LM-555 device.

be used, although for this test the 2N3906 device was used. The zener diode is a 5.6 VDC unit. Note that the output is taken from pins 6 and 7, rather than the regular chip output, pin no. 3, which is not used.

The circuit as shown is a one-shot multivibrator. Triggering occurs in the 555 when pin no. 2 is brought to a potential less than 2/3 the supply potential. When a pulse is applied to pin no. 2 through differentiating network R1C1, the device will trigger because the negative-going slope meets the triggering criteria. To make an astable sawtooth multivibrator drive the input of this circuit with either a square wave or pulse train that produces at least one pulse for each required sawtooth. Being a non-retriggerable monostable multivibrator, the circuit of **Figure 5-5b** will ignore subsequent trigger pulses during the one-shot's refractory period.

Chapter 6

Other IC Timers

The LM-555 is probably the most common form of IC timer on the market, but it is far from the only device available. In this chapter we will take a brief look at several other versions, including the XR-2240, LM-122/LM-322, and LM-2905/LM-3905 timers. There are others on the market, but these are well known and very popular.

XR-2240 IC Timer

The XR-2240 (also designated 8240) IC timer (**Figure 6-1**) is based on the same circuit concept as the 555 device; i.e., a window comparator that sets and resets a control flip-flop. The timebase circuit must receive its power from outside the chip, so in normal operation a 20 kohm resistor is connected between the regulator output (pin no. 15) and timebase output (pin no. 14). This feature allows external timebase circuits to be used.

The timebase section of the XR-2240 is basically the 555-style timer modified to remove the constant 1.1 from the timing equation (see **Equation 6-1**). The reason for the constant in the 555 is that the resistors in the internal voltage divider that biases the comparators are equal. But in the XR-2240 non-equal resistances are used, making the reference levels 0.27(V+) and 0.73(V+) instead of 0.33(V+) and 0.67(V+) as used in the 555. This change was made in order to simplify the timing equation to:

$$T = R \times C \quad \text{eq. (6-1)}$$

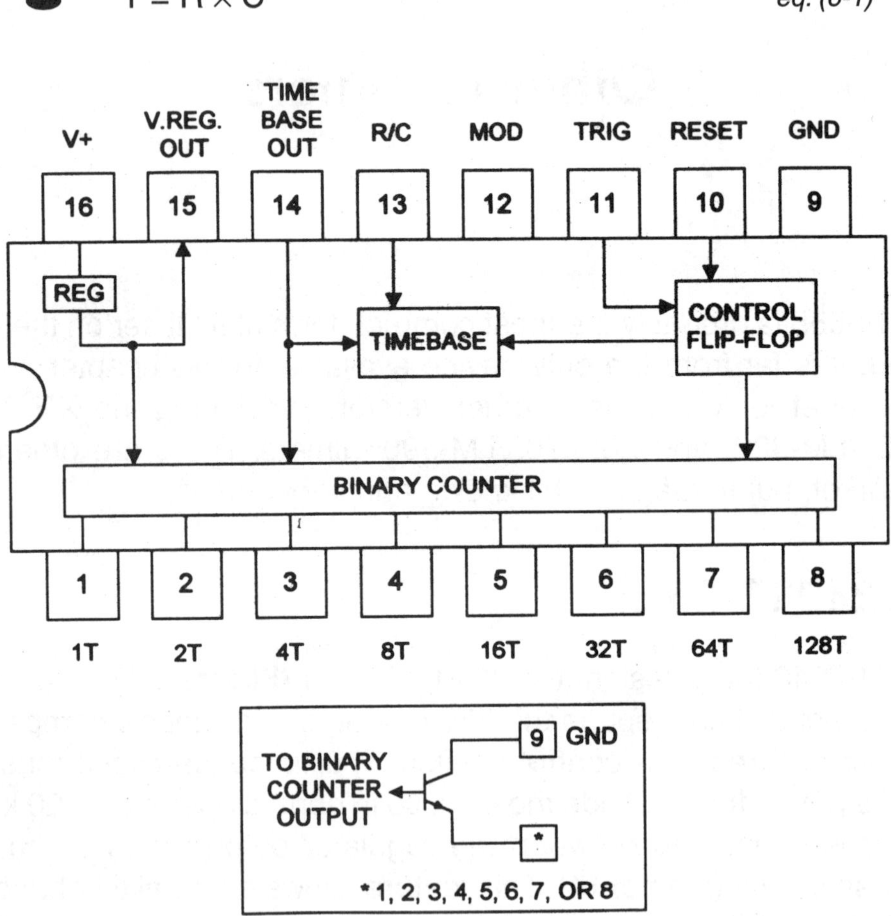

Figure 6-1. The XR-2240 IC timer.

The values of the timing components are 1 kohm £ R £ 10 megohms, and 0.05 µF £ C £ 1000 µF. The XR-2240 is packaged in a 16-pin DIP case, and will operate over the range +4.5 VDC to +18 VDC.

The XR-2240 differs from the 555 in that the trigger and reset pulses are positive-going rather than negative-going. The minimum amplitude of the trigger and reset pulses is approximately two PN junction voltage drops, or about 1.4 VDC. As a practical matter, however, a minimum 3 VDC level is recommended in order to guard against trigger failures caused by negative noise impulses.

There are other differences between the 555 device and the XR-2240. The XR-2240 contains an eight-bit binary counter with open-collector NPN transistor outputs (see inset to **Figure 6-1**). If these outputs are wired in the logical-OR configuration, then times from 1T to 255T can be programmed, where *T* is defined by **Equation 6-1** above. Each output is connected to V+ through a 10 kohm pull-up resistor. The use of the open-collector configuration makes the output active-LOW. The combination of the large range of timer RC components and an eight-bit binary counter makes it possible to create very long duration timer circuits that are essentially as stable as the RC network.

It is also possible to use the binary outputs independently of each other if each is supplied with its own 10 kohm pull-up resistor. Examples of such applications include binary address sequences for digital circuits, and oscillators with base-2 related synchronized output frequencies.

Other, less common, versions of the timer include the XR-2250 (or 8250), which offers Binary Coded Decimal (BCD) outputs instead of binary, and the XR-2260, which uses BCD outputs but limits the most significant digit to "6."

Figure 6-2 shows the circuit for both the monostable and astable multivibrator configurations for the XR-2240. One interesting design feature of the XR-2240 is that the sole difference between the monostable and astable modes is a single feedback resistor (R3) that automatically re-triggers the XR-2240 at the end of each output cycle.

The timer is set into operation by applying a positive-going pulse to the trigger input (pin no. 11). This pulse is routed to the control logic, and performs several jobs simultaneously: resetting the binary counter to 00000000_2, driving all outputs HIGH and enabling the timebase circuit. As was true with the 555 timer, the XR-2240 works by charging capacitor C1 through resistor R1 from positive voltage source V+.

One purpose for the open-collector configuration is that different multiples of the basic timer duration can be programmed by connecting the required outputs together in a wire-OR configuration. For example, if a 57-second timer is needed, it is possible to use a 1 megohm resistor and a 1 µF capacitor (T = 1 second), and then connect together the 1T, 8T, 16T and 32T outputs: (1 + 8 + 16 + 32)T = 57T. In that circuit pins 1, 4, 5 and 6 will be connected together, and to V+ through a single 10 kohm resistor. The output will remain LOW for 57 seconds following triggering, and then return to the HIGH state.

The XR-2240 is considered by some to be more flexible than the 555 because the duration (monostable mode) or period (astable mode) can be set by either the RC timing network or through selection of which outputs are wired together at the output of the circuit.

Synchronization to an external timebase, or modulation of the pulse width, is possible by manipulation of the modulation input (pin no. 12). In normal operation, the modulation input is bypassed to ground through a 0.01 µF capacitor so that noise signals will not disrupt operation of the device. A voltage applied to pin no. 12 will modulate the pulse width of the timebase output signal. This modulating voltage should be between +2 and +5V for a change factor of 0.4 to 2.25.

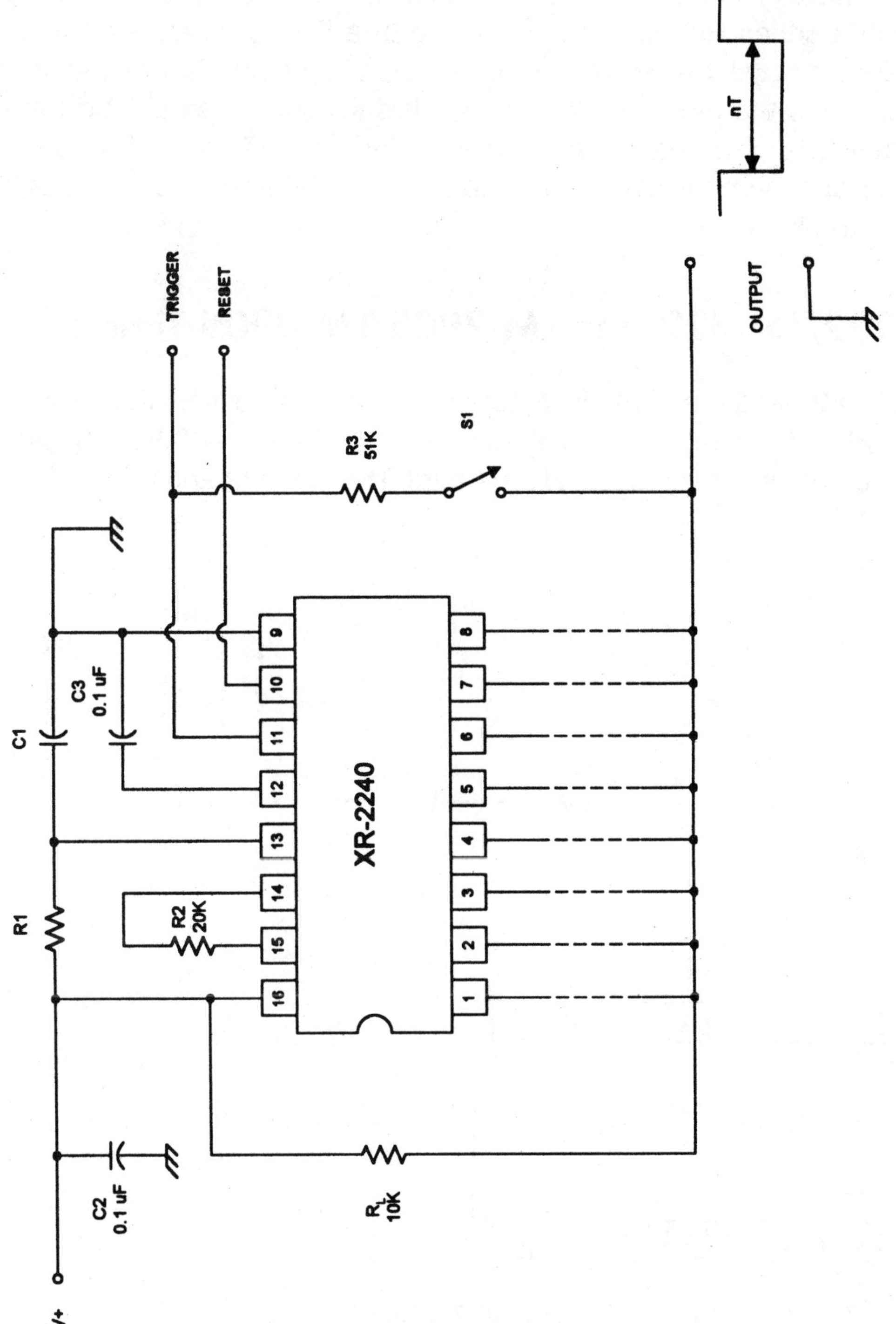

Figure 6-2. Astable and monostable circuit for XR-2240.

Synchronization of the XR-2240 to an external timebase is accomplished through a series RC network consisting of a 5.1 kohm resistor and a 0.1 µF capacitor (see **Figure 6-3**) connected to pin no. 12. This network differentiates the input pulse. The synchronization pulse should have an amplitude of at least +3V, and a period of 0.3T to 0.8T. Another method of synchronization is to use an external timebase connected directly to pin no. 14.

LM-122/LM-322 and LM-2905/LM-3905 Timers

The LM-X22 and LM-X905 IC timers are precision devices that will operate with DC power supply voltages of +4.5 VDC to +40 VDC to produce durations of microseconds to hours. The LM-122 and LM-322 de-

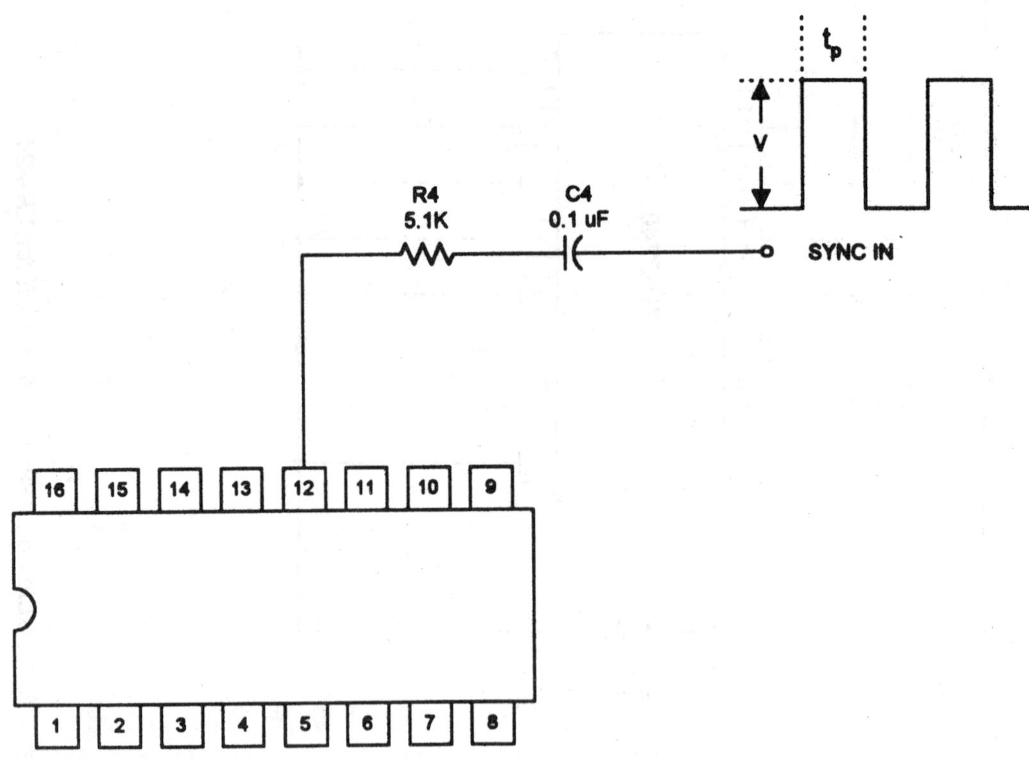

Figure 6-3. Synchronized oscillation of XR-2240 astable.

vice package is shown in **Figure 6-4a**, while that of the LM-2905 and 3905 is shown in **Figure 6-4b**. The LM-2905/3905 devices are identical to the LM-122/322 except that the BOOST and V_{adj} pins are not available. The LM-122 and LM-2905 operate over the temperature range -55°C to +125°C, while the LM-322 and LM-3905 operates over 0°C to +70°C.

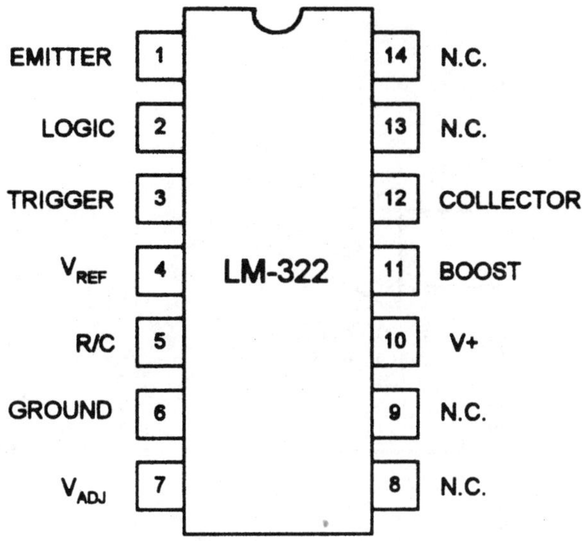

Figure 6-4a. LM-322 timer pin-outs.

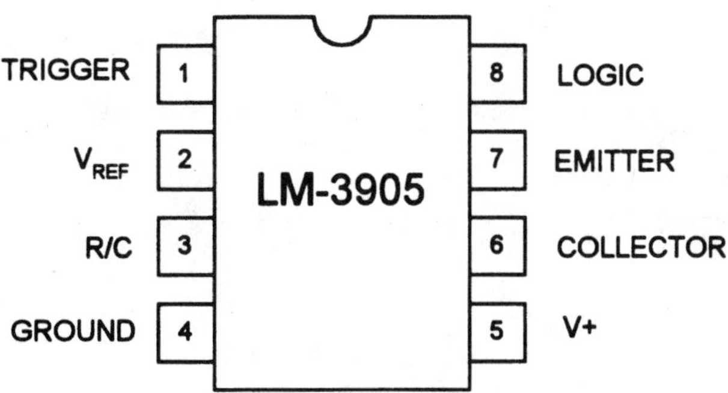

Figure 6-4b. LM-3905 timer pin-outs.

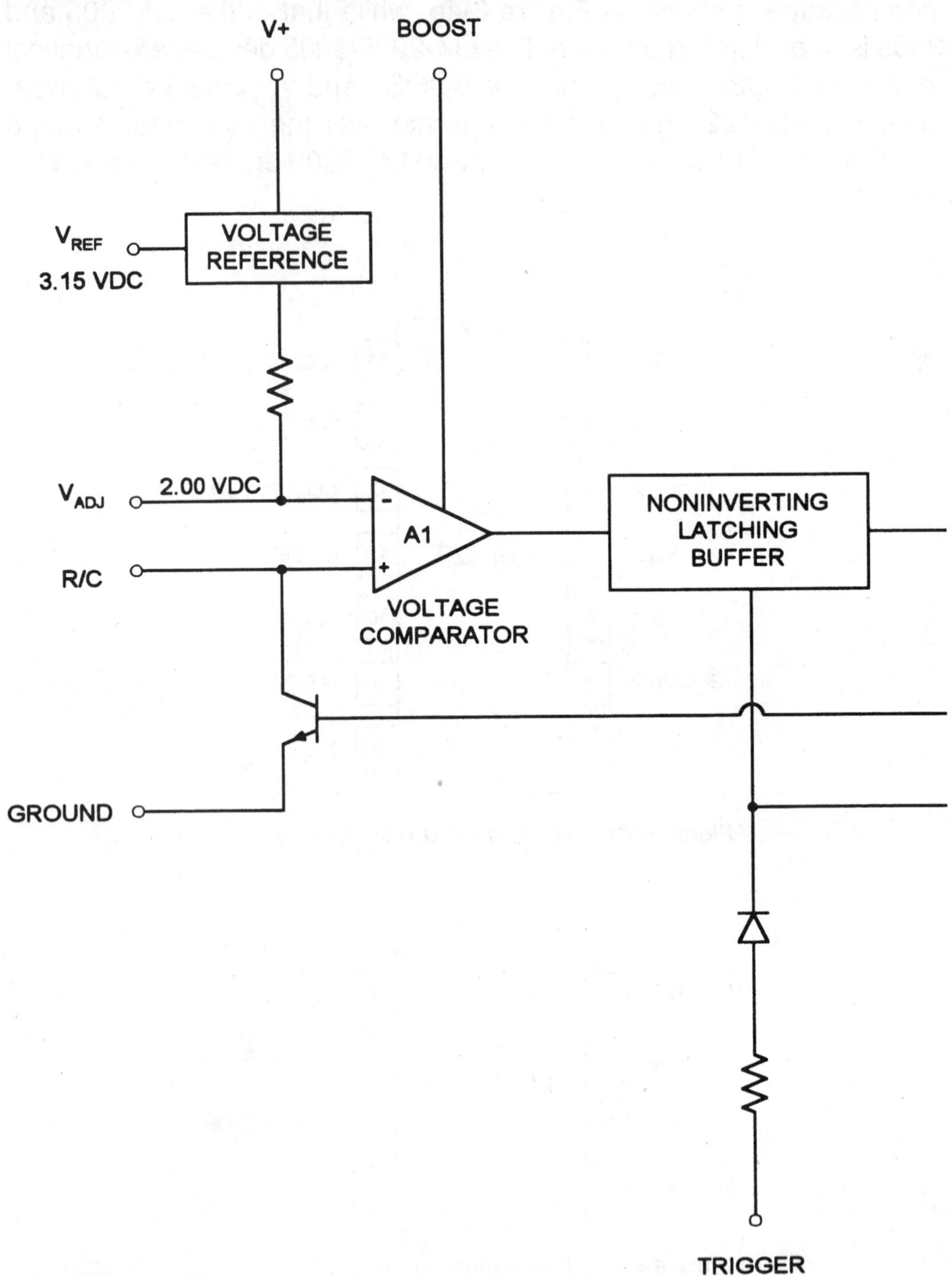

Figure 6-4c. Internal circuitry.

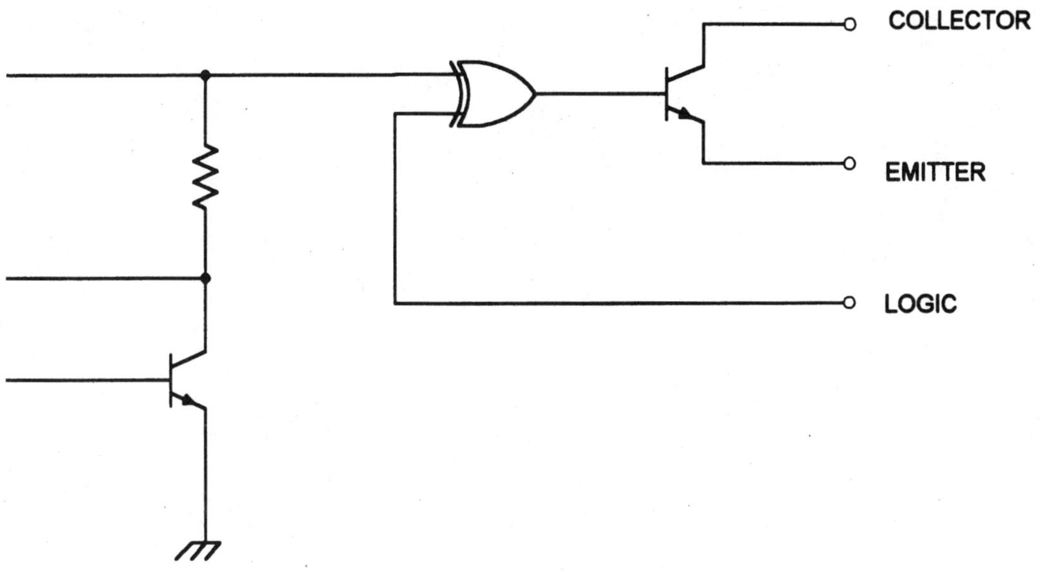
COLLECTOR
EMITTER
LOGIC

The internal circuitry of these timers is shown in **Figure 6-4c**. The RC timing network is monitored by the noninverting input of a voltage comparator (A1); the inverting input is biased to +2.00 VDC. The output of the comparator is routed to a noninverting latching buffer, which in turn drives an Exclusive-OR (XOR) gate. The alternate input of the XOR gate is connected to the LOGIC pin on the device. The output of these devices is a floating emitter and floating collector transistor. Both ground referenced and floating loads, at potentials up to 40-VDC, can be accommodated with this arrangement. The V_{adj} pin can be used in the LM-122 and LM-322 devices to vary the timing ratio up to 50:1 by using an external voltage. The circuit for the basic timer is shown in **Figure 6-5**. The output duration is set by T = R1C1.

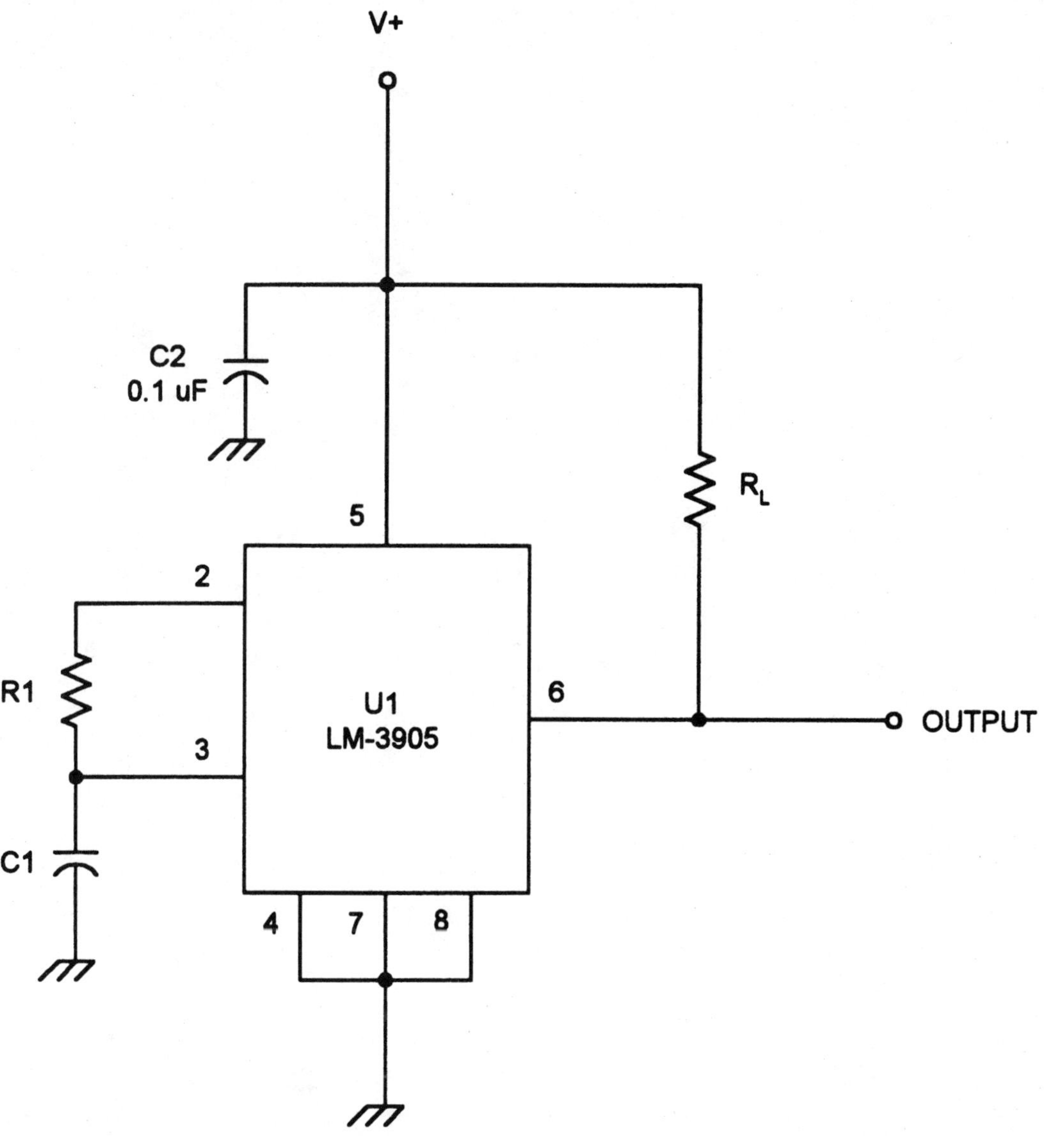

Figure 6-5. The LM-3905 circuit.

Chapter 7

Operational Amplifier Timer Circuits

The operational amplifier (op-amp) is a commonly available integrated circuit that has a large variety of uses, and some of those uses are in timer circuits. In this chapter we will take a look at the op-amp circuits that are usable for timer applications, notably the monostable multivibrator or one-shot circuit. But first, let's take a look at some basic op-amp theory.

The Operational Amplifier

The operational amplifier was invented in the late 1940s for solving mathematical operations (hence the name) in analog computers. Modern devices are no longer made of vacuum tubes as were the 1949 devices. They are in the form of some very sophisticated integrated circuits. These devices, unlike their earlier counterparts, actually come close to the ideal properties of the theoretical op-amp.

The symbol for the operational amplifier is shown in **Figure 7-1**. The pin-outs shown are not found on all devices, but are considered industry standard so will be found on most devices. Note that there are two DC power supplies, but no ground terminal (ground is established at the common point of the two DC power supplies). The V+ DC power supply (pin no. 7) is positive with respect to common, and the V- DC power supply (pin no. 4) is negative with respect to common (or ground). The output terminal of the device (pin no. 6) is single-ended, so the signal is taken between this terminal and ground.

There are also two input terminals (pins 2 and 3). Having two inputs is not part of the required properties of an op-amp, but only a few commercial examples have only one op-amp. The -IN terminal is called the inverting input, while the +IN terminal is called the noninverting input. The -IN terminal (pin no. 2) and +IN terminal (pin no. 3) look into the same impedance values, and offer the same gain. The difference is that the -IN input produces an output signal that is 180 degrees out of

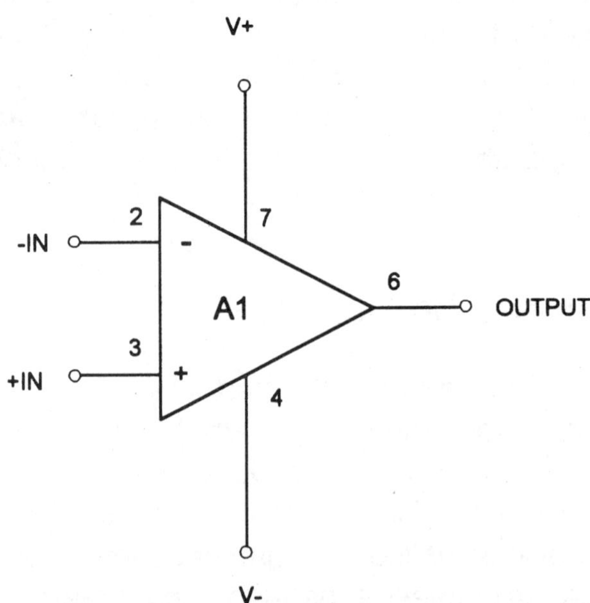

Figure 7-1. Operational amplifier symbol.

phase with the input signal, and the +IN input produces an output signal that is in phase (0° or 360°). The output signal (V_O) is proportional to the input voltage and the voltage gain (A_V):

For inverting input (-IN):

- $V_O = -V_{IN} \times A_V$ *eq. (7-1)*

For noninverting input (+IN):

- $V_O = V_{IN} \times A_V$ *eq. (7-2)*

When the same signal is applied to both -IN and +IN the output is zero. In **Equation 7-3** below the gain is the differential voltage gain (A_{VD}), and the output voltage is:

For inverting input (-IN):

- $V_O = A_V \times (V_{+IN} - V_{-IN})$ *eq. (7-3)*

Examples of common op-amps are shown in **Figure 7-2**. The device shown in **Figure 7-2a** uses the same pin-outs as the device in **Figure 7-1**. It is the usual but not universal pin-outs of single op-amp devices packaged in an 8-pin mini-DIP package (e.g. 741, CA-3140). The version shown in **Figure 7-2b** is also in an 8-pin mini-DIP package, but contains two operational amplifiers. These pin-outs are common to LM-1458, CA-3240 and so forth.

The single op-amp of **Figure 7-2a** has the following pins (the same pin definitions, except the offset null, are also found on the dual device):

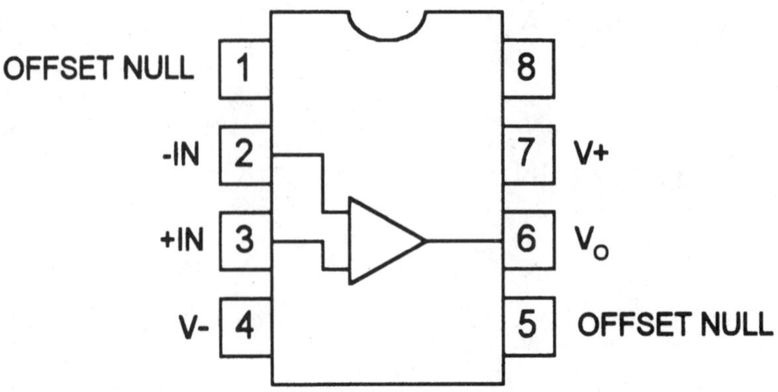

Figure 7-2a. Industry standard single op-amp pin-outs.

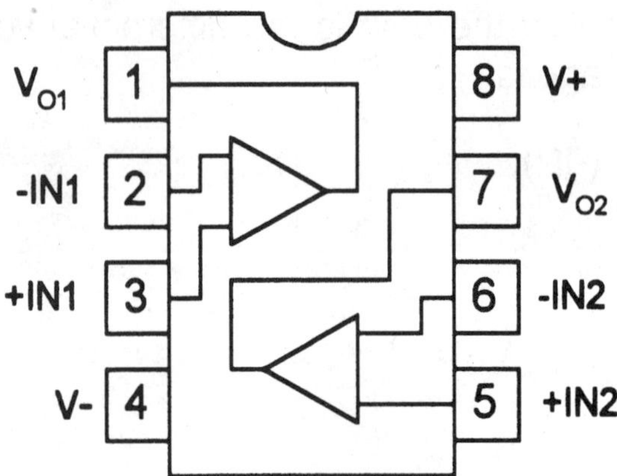

Figure 7-2b. Dual device pin-outs.

<u>**Inverting Input (-IN), Pin No. 2**</u>. The output signals produced from this input are 180 degrees out of phase with the input signal applied to -IN.

Noninverting Input (+IN), Pin No. 3. Output signals are in phase with signals applied to the +IN input terminal.

Output, Pin No. 6. On most op-amps, the 741 included, the output is single-ended. This term means that output signals are taken between this terminal and the power supply common (see **Figure 7-3**). The output of the 741 is said to be short-circuit-proof because it can be shorted to common indefinitely without damage to the IC.

V+ DC Power Supply, Pin No. 7. The positive DC power supply terminal.

V- DC Power Supply, Pin No. 4. The negative DC power supply terminal.

Offset Null, Pins 1 and 5. These two terminals are used to accommodate external circuitry that compensates for offset error voltages.

Op-Amp Signal Designations

Also shown in **Figure 7-3** are the output signals. The output voltage is designated V_O, and is taken between the DC power supply common (which may be grounded or not). The signal applied to the inverting input (-IN) is V1, and that applied to the noninverting input (+IN) is V2. Each of these signals is single-ended with respect to common, but taken together form a differential signal (V_D):

- $$V_D = V2 - V1$$ *eq. (7-4)*

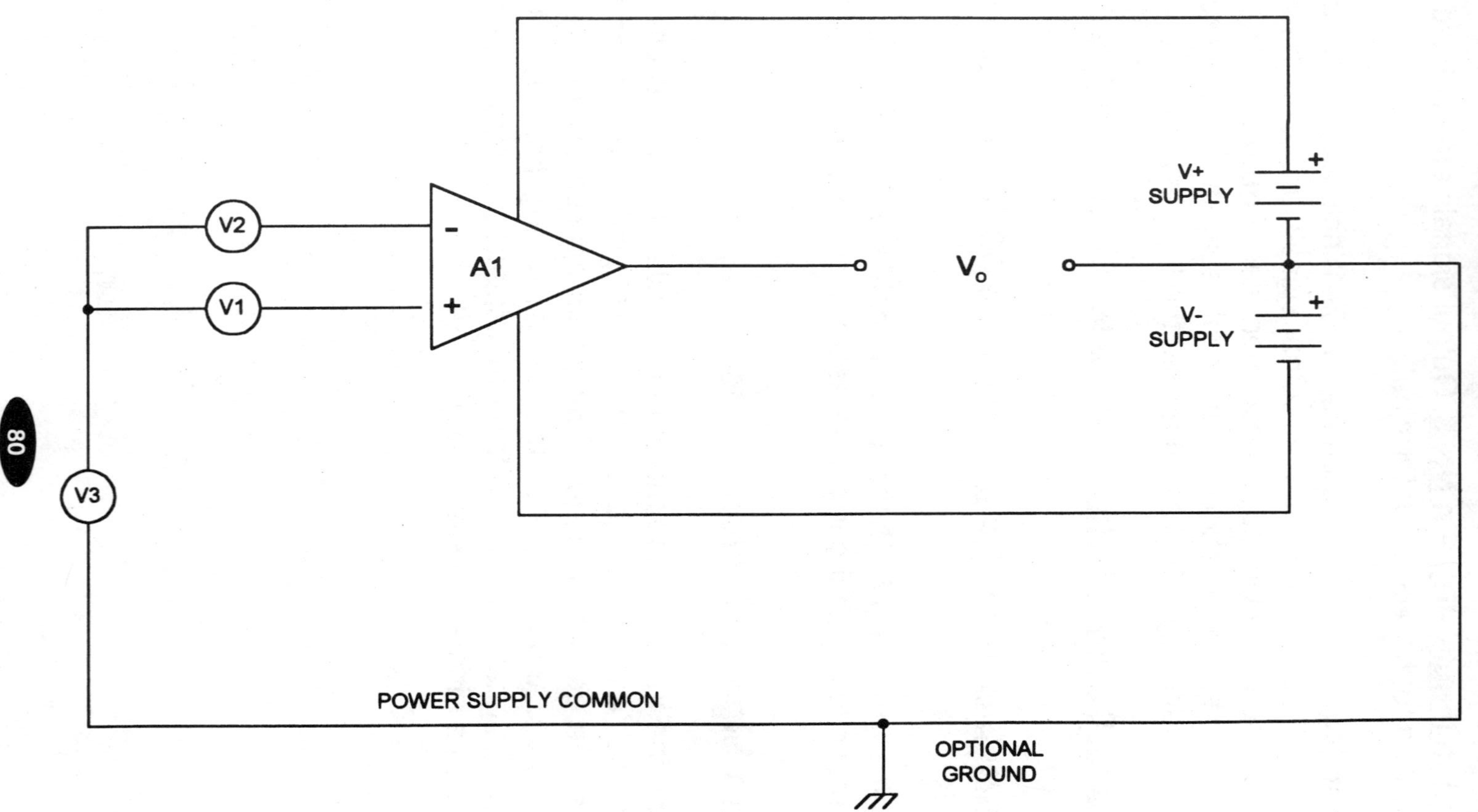

Figure 7-3. Signal and power supply configuration for an operational amplifier.

The Ideal Operational Amplifier

When you study any type of electronic device it is wise to start with an ideal model of that device, and then proceed to less ideal practical devices. In some cases, the practical and ideal devices are so far apart that you might wonder at the wisdom of this approach. But IC operational amplifiers, even low-cost products, so nearly approximate the ideal op-amp of textbooks that the equations actually work. The ideal model analysis method thus becomes extremely useful for understanding the technology, learning to design new circuits, or figuring out how an unfamiliar circuit works.

Properties of the Ideal Operational Amplifier

The ideal op-amp is characterized by seven properties. From this short list of properties we can deduce circuit operation and design equations. Also, the list gives us a basis for examining non-ideal operational amplifiers and their defects (plus solutions to the problems caused by those defects). The basic properties of the op-amp are:

1. Infinite open-loop voltage gain
2. Infinite input impedance
3. Zero output impedance
4. Zero noise contribution
5. Zero DC output offset
6. Infinite bandwidth
7. Both differential inputs stick together

Let's take a look at these properties to determine what they mean in practical terms. You will find that some cheap op-amps only approximate some of these ideals, while for others on the list the approximation is extremely good.

Property No. 1 — Infinite Open-Loop Gain (A_{vol}). The open-loop gain of any amplifier is its gain without either negative or positive feedback. By definition, negative feedback is a signal fed back to the input 180 degrees out of phase. In operational amplifier terms this means feedback between the output and the inverting input.

Negative feedback has the effect of reducing the open-loop gain (A_{vol}) by a factor (β) that depends on the transfer function and properties of the feedback network. **Figure 7-4** shows the basic configuration for any negative feedback amplifier. The transfer equation for any circuit is the ratio of the output function and the input function. The transfer function of a voltage amplifier is, therefore, $A_{vol} = V_o/V_{in}$. In **Figure 7-4** the term A_{vol} represents the gain of the amplifier element only, i.e., the gain with the feedback network disconnected. The overall transfer function of this circuit, i.e., with both amplifier element and feedback network (β) in the loop, is defined as:

$$A_v = \frac{A_{vol}}{1 + A_{vol}\beta} \qquad \text{eq. (7-5)}$$

Where:

A_v is the closed-loop gain

A_{vol} is the open-loop gain

β is the transfer equation for the isolated feedback network

In the ideal op-amp, A_{vol} is infinite, so (as the math reveals) the voltage gain is a function only of the characteristics of the feedback network. This is an important point, so remember it well! In real op-amps, the value of the open-loop gain is not quite infinite, but it is extremely high. Typical values range from 20,000 in low-grade consumer audio models to more than 2,000,000 in premium units (in common 741 devices it is typically 200,000 to 300,000).

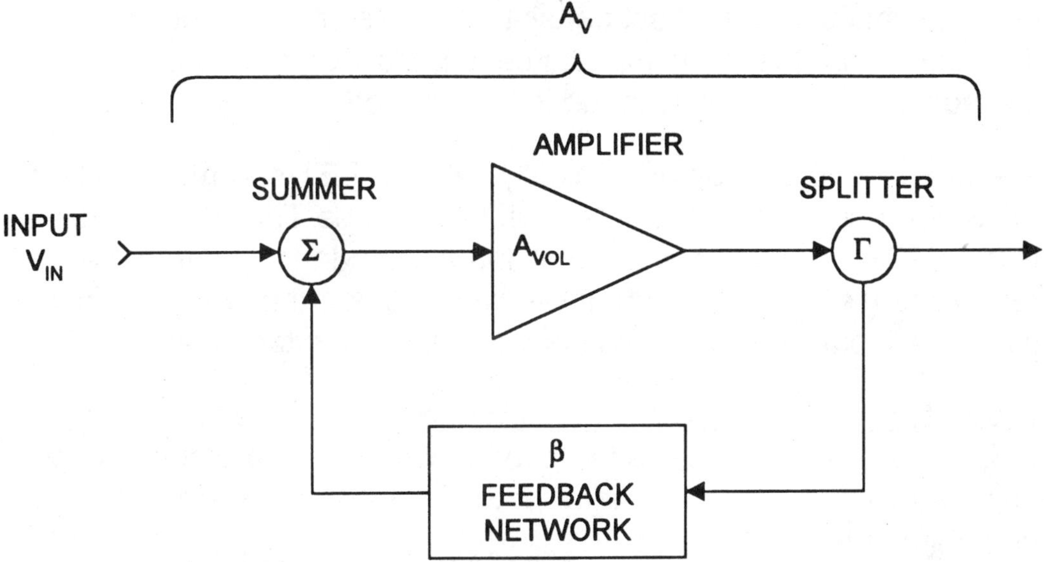

Figure 7-4. Block diagram of a feedback amplifier.

Property No. 2 — Infinite Input Impedance. This property implies that the op-amp input will not load the signal source. The input impedance of any amplifier is defined as the ratio of the input voltage the input current: $Z_{in} = V_{in}/I_{in}$. When the input impedance is infinite, therefore, we must assume that the input current is zero. Thus, an important implication of this property is that the operational amplifier inputs neither sink nor source current. In other words, it will neither supply current to an external circuit, nor accept current from an external circuit. We will depend upon an implication of this property ($I_{in} = 0$) to perform the simplified circuit analysis used in this chapter to determine the gain equations.

Real operational amplifiers have some finite input current other than zero. In low-grade devices this current can be substantial (e.g., >1 mA), and will cause a large output offset voltage error in medium and high gain circuits. The primary source of this current is the base bias currents from the NPN and PNP bipolar transistors used in the input circuits. Certain premium grade op-amps that feature bipolar inputs reduce this current to nanoamperes (10^{-9} A) or picoamperes (10^{-12} A). In op-amps that use field effect transistors (FETs) in the input circuits, on the other hand, the input impedance is quite high due to the very low leakage currents normally found in FET devices.

The JFET input devices are typically called BiFET op-amps, while the MOSFET input models are called BiMOS devices. The CA-3140 device is a BiMOS op-amp in which the input impedance approaches 1.5 terraohms (i.e., 1.5×10^{12} ohms) — which is near enough to infinite to make the input circuits of those devices approach the ideal.

Property No. 3 — Zero Output Impedance. A voltage amplifier (of which class, the op-amp is a member) ideally has a zero output impedance. All real voltage amplifiers, however, have a non-zero (but low) output impedance.

Real operational amplifiers do not have a zero output impedance. The actual value is typically less than 100 ohms, with many being in the neighborhood of 30 ohms. Thus, for typical devices the operational amplifier output can be treated as if it were ideal.

A rule of thumb used by designers is to set the input resistance of any circuit that is driven by a non-ideal voltage source output to at least ten times the previous stage output impedance.

Property No. 4 — Zero Noise Contribution. All electronic circuits, even simple resistor networks, produce noise signal at temperatures above absolute zero. A resistor creates noise due to the thermal movement of electrons in its internal resistance element material. In the ideal operational amplifier, zero noise voltage is produced internally. Thus, any noise in the output signal must have been present in the input signal as well. Except for amplification, the output noise voltage will be exactly the same as the input noise voltage. In other words, the op-amp contributed nothing extra to the output noise. This is one area where practical devices depart quite a bit from the ideal. Practical op-amps do not approximate the ideal, except for certain higher cost premium low-noise models.

Amplifiers use semiconductor devices that create not merely resistive noise (as described above), but also create other noise of their own. There are a number of internal noise sources in semiconductor devices, and any good text on transistor theory will give you more information on them. For present purposes, however, assume that the noise contribution of the op-amp can be considerable in low signal level situations. Premium op-amps are available in which the noise contribution is very low, and these devices are usually advertised as premium low-noise types. Others, such as the metal can CA-3140 device, will offer relatively low noise performance when the DC supply voltages are limited to ±5 VDC, and the metal package of the op-amp is fitted with a flexible "TO-5" style heatsink.

Property No. 5 — Zero Output Offset. The output offset voltage of any amplifier is the output voltage that exists when it should be zero. The voltage amplifier sees a zero input voltage when the inputs are both grounded. This connection should produce a zero output voltage. If the output voltage is non-zero, then there is an output offset voltage present. In the ideal op-amp, this offset voltage is zero volts — but real op-amps exhibit at least some amount of output offset voltage. In the real IC operational amplifier the output offset voltage is non-zero, although it can be quite low in some cases.

Property No. 6 — Infinite Bandwidth. The ideal op-amp will amplify all signals from DC to the highest AC frequencies. In real op-amps, however, the bandwidth is sharply limited. There is a specification called the gain-bandwidth (G-B) product, which is symbolized by F_t. This specification is the frequency at which the voltage gain drops to unity (1). The maximum available gain at any frequency is found by dividing the maximum required frequency into the G-B product. If the value of F_t is not sufficiently high, then the circuit will not behave in classical op-amp fashion at some higher frequencies within the range of interest.

Some op-amps have G-B products in the 10 to 20 MHz range. Others, on the other hand, are quite limited. A few special devices have G-B products up to 600 MHz. The 741-family of devices is very limited, such that the device will perform as an op-amp only to frequencies of a few kilohertz. Above that range, the gain drops off considerably. But in return for this apparent limitation, we obtain nearly unconditional stability; such op-amps are said to be frequency compensated. It is the frequency compensation of those devices which both reduces the G-B product and provides the inherent stability. Non-compensated op-amps will yield wider frequency response, but only at the expense of a tendency to oscillate. Those op-amps may spontaneously oscillate without any special encouragement if certain precautions are not taken in the circuit design.

Property No. 7 — Differential Inputs Stick Together. Most operational amplifiers have two inputs: an inverting (-IN) input and an noninverting (+IN) input. *Sticking together* means that a voltage applied to one of these inputs also appears at the other input. This voltage is real — it is not merely some theoretical device used to evaluate circuits. If you apply a voltage to, say, the inverting input, and then connect a voltmeter between the noninverting input and the power supply common, then the voltmeter will read the same potential on the noninverting as it did on the inverting input. The implication of this property is that both inputs must be treated the same mathematically. This fact will make itself felt when we discuss the concept of virtual as opposed to actual grounds, and again when we deal with the noninverting follower circuit configuration.

The inverting follower produces an output signal that is 180 degrees out of phase with its input signal. The noninverting follower, as you might expect, produces an output signal that is in phase with its input signal. Almost all other operational amplifier circuits are variations on either inverting or noninverting follower circuits. Understanding these two configurations will allow you to understand, and either design or modify, a wide variety of different circuits using IC operational amplifiers.

Standard Circuit Configurations

There are two basic circuit configurations of interest to us in this chapter: inverting and noninverting. There are other configurations, but these define what we need to discuss for timers (and they are also the basic forms).

The circuit for the inverting follower is shown in **Figure 7-5**. The noninverting input is grounded, and signal is applied to the inverting input through input resistor R_{IN}. This circuit produces an output signal that is 180 degrees out of phase with the input signal. There is also a feedback resistor, R_F, between the output terminal and the inverting input. Taken together R_{IN} and R_F form the feedback network that sets the voltage gain. The output voltage of this circuit is:

$$V_O = -V_{IN} \frac{R_F}{R_{IN}} \qquad \text{eq. (7-6)}$$

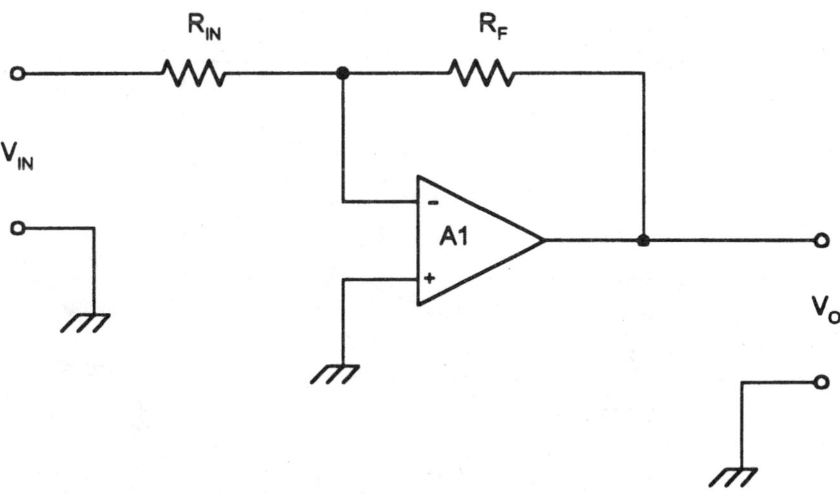

Figure 7-5. An inverting follower circuit.

Because the noninverting input is grounded, and thus at zero volts potential, the inverting input is also at the ground potential (see ideal properties above). This means that the input impedance of this circuit is limited to the value of R_{IN}.

The circuit for the noninverting follower is shown in **Figure 7-6**. In this circuit, the signal V_{IN} is applied to the noninverting input. The output signal is therefore in phase with the input signal. Note that we still use the same form of feedback network, but the input end of the input resistor is grounded.

The output voltage of this circuit is given by:

$$V_O = V_{IN}\left[\frac{R_F}{R_{IN}} + 1\right] \quad \text{eq. (7-7)}$$

The input impedance of this circuit is very high, and is essentially the natural input impedance of the operational amplifier. The noninverting follower circuit of **Figure 7-6** is used in the following monostable multivibrator circuit.

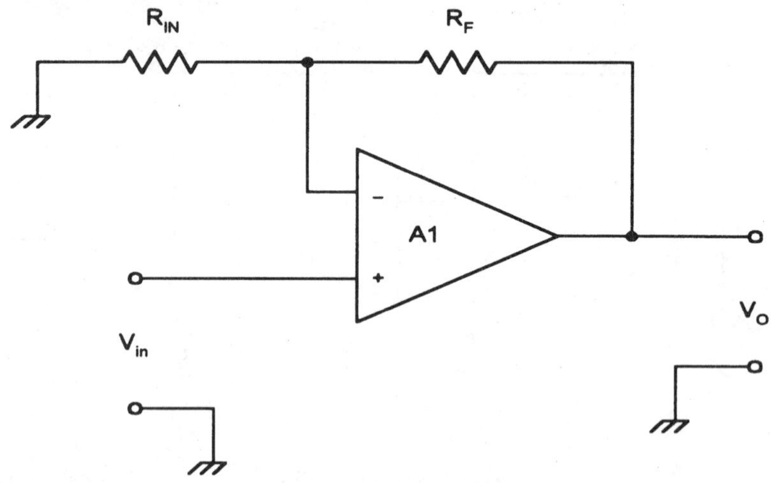

Figure 7-6. Noninverting follower circuit.

Monostable Multivibrator Circuits

Recall that the MMV has two permissible output states (HIGH and LOW), but only one of them is stable. The MMV produces one output pulse in response to an input trigger signal (**Figure 7-7**). The output pulse (V_o) has a time duration, *T*, in which the output is in the quasistable state. The MMV is also known under several alternate names: one-shot, pulse generator, and pulse stretcher. The latter name derives from the fact that the output pulse duration, *T*, is longer than the trigger pulse duration ($T > T_c$).

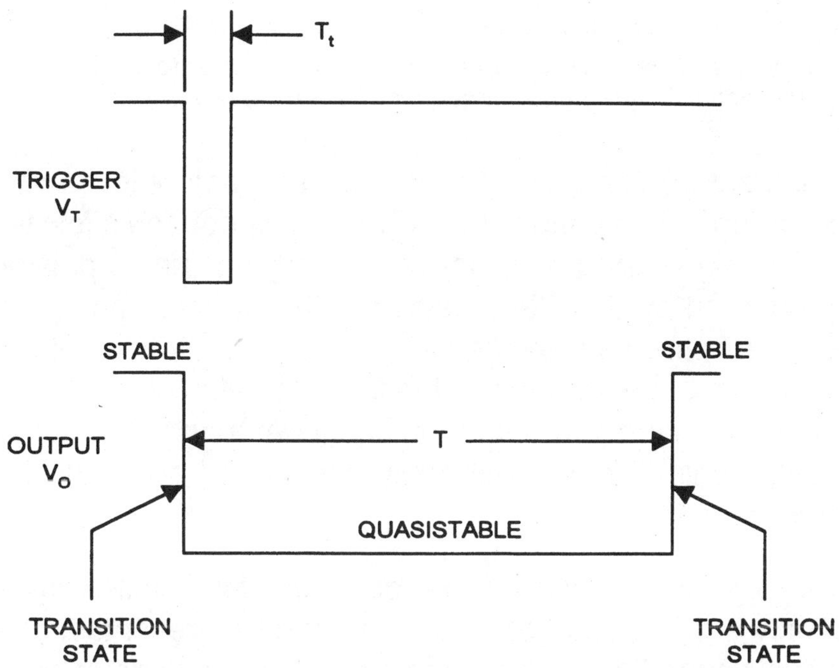

Figure 7-7. Timing for a monostable multivibrator.

Monostable multivibrators find a wide variety of applications in electronic circuits. Besides the pulse stretcher mentioned above, the MMV also serves to lock out unwanted pulses. **Figure 7-8** shows that the output responds to only the first trigger pulse. The next two pulses occur during the active time, *T*, so are ignored. Such an MMV is said to be

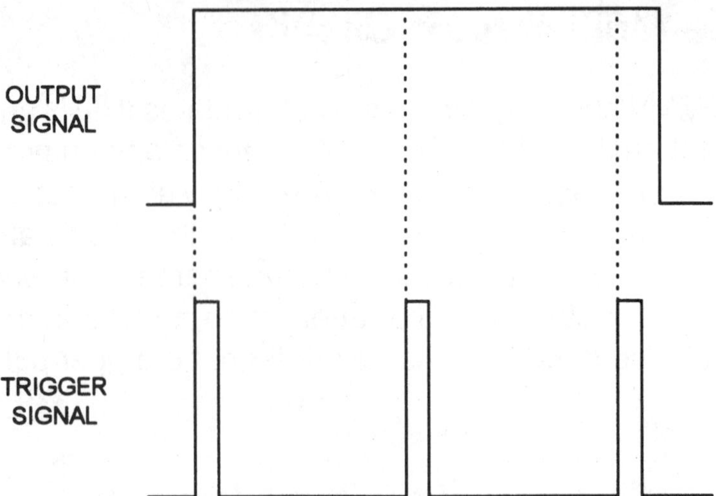

Figure 7-8. Response of a monostable multivibrator to succeeding trigger pulses that occur before time-out.

non-retriggerable. A common application of this feature is in switch contact debouncing. All mechanical switch contacts bounce a few times on closure, creating a short exponentially decaying run of pulses. If an MMV is triggered by the first pulse from the switch, and if the MMV remains quasi-active long enough for the bouncing to die out, then the MMV output signal becomes the debounced switch closure. The main requirement is that the MMV duration be longer than the switch contact bounce pulse train; 5 mS is generally considered adequate for most switch types.

The range of MMV applications is too broad for detailed discussion here, so only a general set of categories can be presented. These include: pulse generation, pulse stretching, contact debouncing, pulse signal clean-up, switching, and synchronization of circuit functions (especially digital).

Figure 7-9a shows the circuit for a non-retriggerable monostable multivibrator based on the operational amplifier. This circuit is based on the voltage comparator circuit, which is an operational amplifier without feedback. When there is no feedback, the effective voltage gain of an

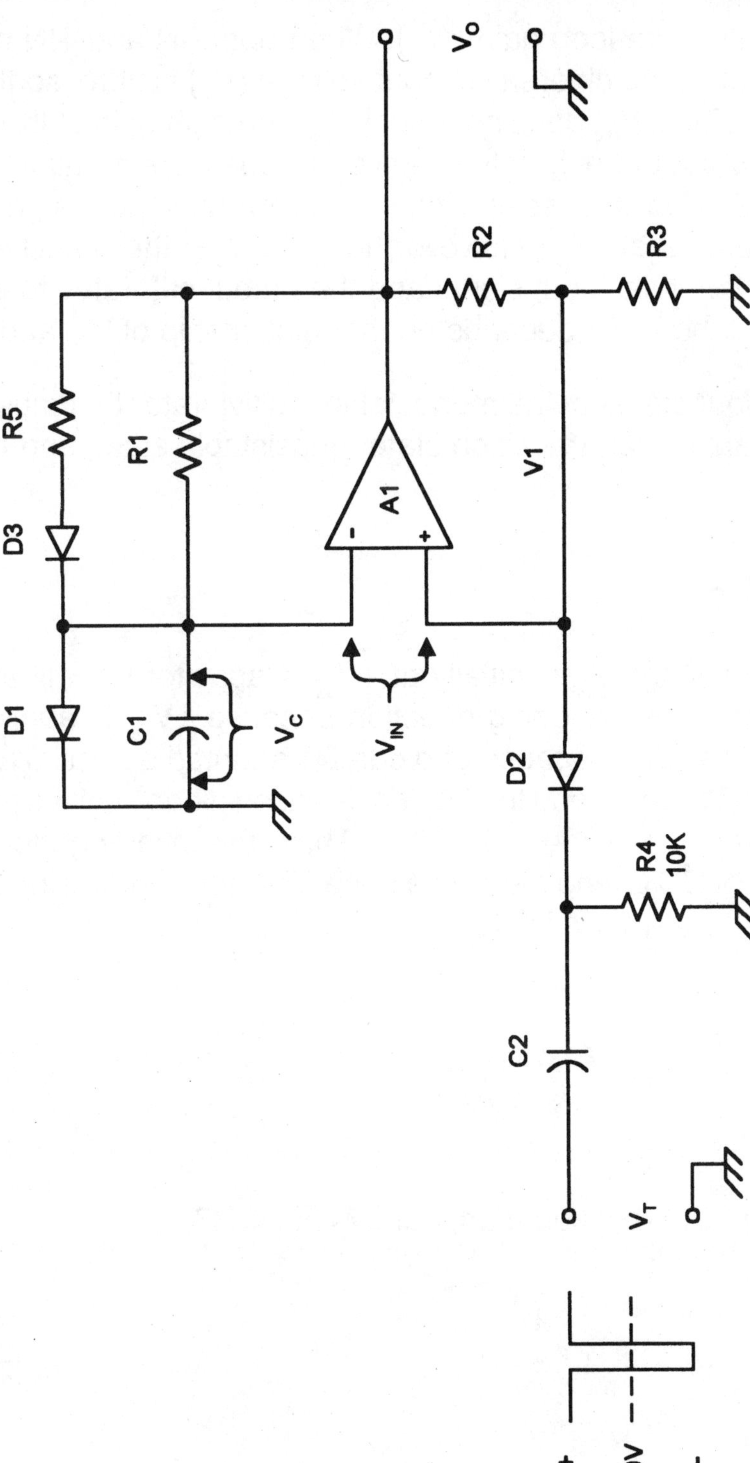

Figure 7-9a. Op-amp monostable multivibrator circuit.

op-amp is its open-loop gain (A_{vol}). When both -IN and +IN are at the same potential, the differential input voltage (V_{id}) is zero, so the output is also zero. But if V_{-IN} does not equal V_{+IN}, the high gain of the amplifier forces the output to either its positive or negative saturation values. If $V_{-IN} > V_{+IN}$, the op-amp sees a positive differential input signal, so the output saturates at $-V_{sat}$. However, if $V_{-IN} < V_{+IN}$, the amplifier sees a negative differential input signal and the output saturates to $+V_{sat}$. The operation of the MMV depends on the relationship of V_{-IN} and V_{+IN}.

There are four states of the monostable multivibrator that must be considered: stable state, transition state, quasistable state, and refractory state.

Stable State

The output voltage V_o is initially at $+V_{sat}$. Capacitor C1 will attempt to charge in the positive-going direction because $+V_{sat}$ is applied to the R1C1 network. But, because of diode D1 shunted across C1, the voltage across C1 is clamped to $+V_{D1}$. For a silicon diode such as the 1N914 or 1N4148 $+V_{D1}$ is about +0.7 VDC. Thus, the inverting input (-IN) is held to +0.7 VDC during the stable state. The noninverting input (+IN) is biased to a level V1, which is:

$$V1 = \frac{R3\,(+V_{sat})}{R2 + R3} \qquad \text{eq. (7-8)}$$

or, in the special (but common) case of R2 = R3:

$$V1 = \frac{+V_{sat}}{2} \qquad \text{eq. (7-9)}$$

The factor R3/(R2 + R3) is often designated by the Greek letter beta (β), so:

$$\beta = \frac{R3}{R2 + R3} \qquad \text{eq. (7-10)}$$

Therefore:

$$V1 = \beta(+V_{sat}) \qquad \text{eq. (7-11)}$$

The amplifier (A1) sees a differential input voltage (V_{id}) of ($V1 - V_{D1}$), or (V1 - 0.7) volts. Using the previous notation:

$$V_{id} = \frac{R3(+V_{sat})}{R2 + R3} \qquad \text{eq. (7-12)}$$

As long as $V1 > V_{D1}$, the amplifier effectively sees a negative DC differential voltage at the inverting input, so (with its high open-loop gain, A_{vol}) the output will remain saturated at $+V_{sat}$. For purposes of this discussion the amplifier is a type 741 operated at DC power supply potentials of ±12 VDC, so V_{sat} will be about ±10V.

Transition State

The input trigger signal (V_t) is applied to the MMV of **Figure 7-9a** through RC network R4C2. The general design rule for this network is that its time-constant should be not more than one-tenth the time-constant of the timing network:

$$R4C2 < \frac{R1C1}{10} \qquad \text{eq. (7-13)}$$

At time T1 (see **Figure 7-9b**), trigger signal V_t makes an abrupt HIGH to LOW transition to a peak value that is less than (V1 - 0.7) volts. Under this condition the polarity of V_{id} is now reversed and the inverting input now sees a positive voltage: ($V1 + V_t - 0.7$) is less than V_{D1}. The output voltage V_o now snaps rapidly to $-V_{sat}$. The fall time of the output signal is dependent upon the slew rate and the open-loop gain of the operational amplifier, A1.

Quasistable State

The output signal from the MMV is the quasistable state shown between T1 and T2 in **Figure 7-9b**. It is called *quasistable* because it does not change over T = T2-T1, but when *T* expires the MMV times-out and V_o reverts to the stable state output voltage ($+V_{Sat}$).

During the quasistable time D1 is reverse-biased, and capacitor C1 discharges from +0.7 VDC to zero and then recharges towards $-V_{sat}$. When $-V_o$ reaches -V1, however, the value of V_{id} crosses zero and that change forces V_o to snap once again to $+V_{sat}$.

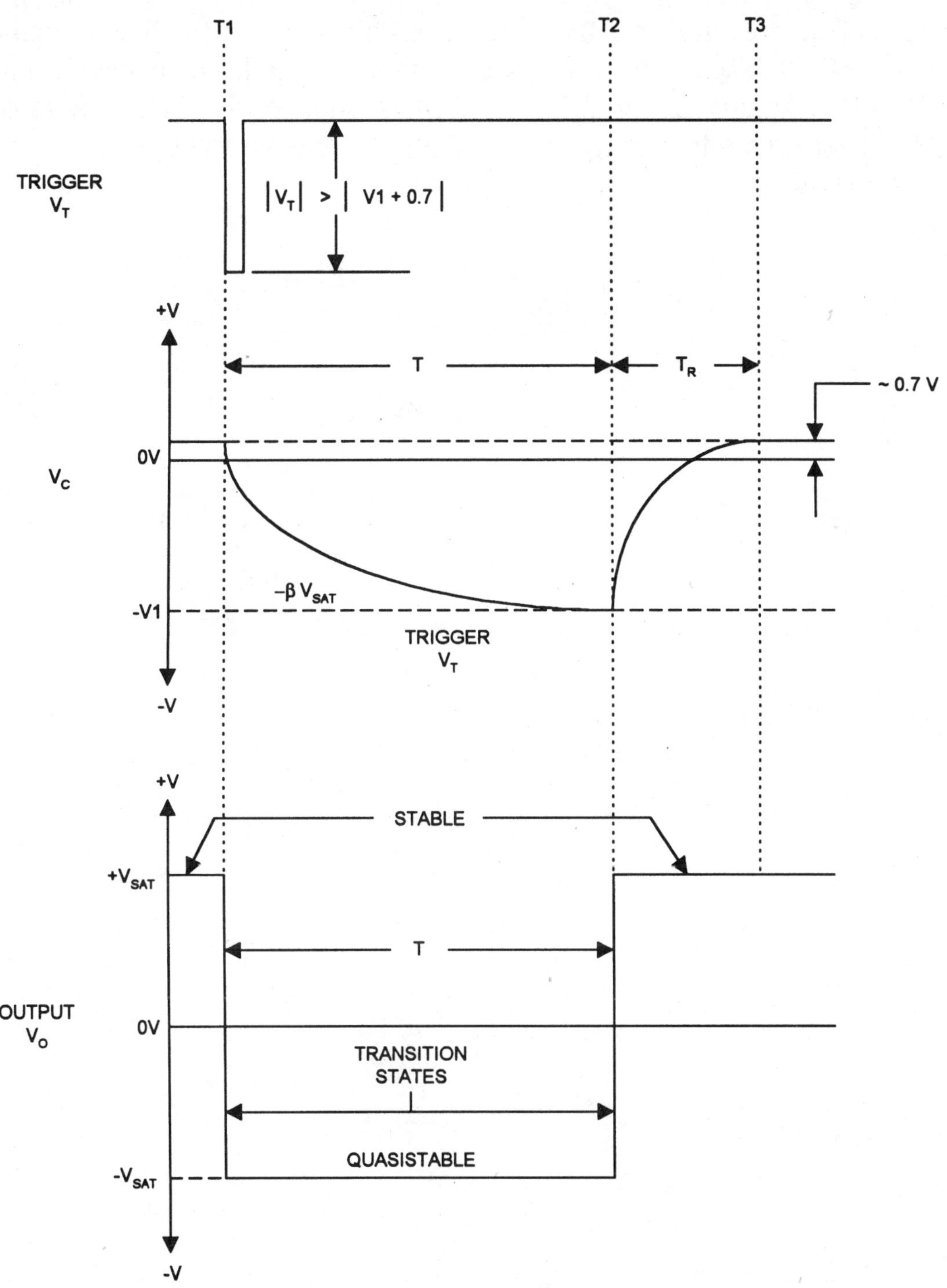

Figure 7-9b. Op-amp monostable multivibrator timing diagram.

Appealing to **Equation 7-8** makes it possible to derive the timing equation for the MMV. The timing capacitor must charge from an initial value (V_{C1}) to a final value (V_{C2}) in time *T*. The question is: "What value of R1C1 will cause the required transitions?" Consider the case R2 = R3 ($V1 = 0.5V_{sat}$):

$$R1C1 = \frac{-T}{LN\left[\frac{V_{sat} - V_{C2}}{V_{sat} - V_{C1}}\right]} \qquad \text{eq. (7-14)}$$

$$R1C1 = \frac{-T}{LN\left[\frac{V_{sat} - ((0.5)(V_{sat} + 0.7))}{V_{sat} - 0.7}\right]} \qquad \text{eq. (7-15)}$$

and, for the case where V_{sat} = 10 VDC:

$$R1C1 = \frac{-T}{LN\left[\frac{10\ Vdc - ((0.5)(10\ Vdc + 0.7))}{10\ Vdc - 0.7}\right]} \qquad \text{eq. (7-16)}$$

$$R1C1 = \frac{-T}{LN\left[\frac{10 - 5.35\ volts}{10\ Vdc - 0.7\ volts}\right]} \qquad \text{eq. (7-17)}$$

$$R1C1 = \frac{-T}{LN\left[\frac{4.65}{9.3}\right]} \qquad \text{eq. (7-18)}$$

- $$R1C1 = \frac{-T}{\text{LN}\ (0.5)}$$ eq. (7-19)

- $$R1C1 = \frac{-T}{-0.69}$$ eq. (7-20)

Thus,

- T = 069 R1 C1

Example

A monostable multivibrator circuit is based on a 741 operational amplifier with an output saturation voltage of ±10V. Calculate the RC time-constant needed to produce a 100 ms output pulse when feedback resistors R2 and R3 are each 10 kohms.

- $$R1C1 = \frac{T}{0.69}$$ eq. (7-21)

- $$R1C1 = \frac{100\ ms\ \_\ \frac{1\ s}{1000\ ms}}{0.69}$$

- $$R1C1 = \frac{0.1}{0.69} = 0.145\ s$$

Equation 7-21 represents the special case in which β = 1/2 (i.e., R2 = R3). Although R2 = R3 may be the usual case for this class of circuit, R2 and R3 might not be equal in other cases. A more generalized expression is:

$$RC = \frac{T}{\mathrm{LN}\left[\dfrac{1 + 0.7V/V_{sat}}{1 - \beta}\right]} \qquad \text{eq. (7-22)}$$

In which,

$$\beta = \frac{R3}{R2 + R3} \qquad \text{eq. (7-23)}$$

When the quasistable state times out, the circuit status returns to the stable state (where it remains dormant until triggered again).

Example

A monostable multivibrator (**Figure 7-9a**) is constructed with R2 = 10 kohms, and R3 = 3.3 kohms. The active device is a 741 operational amplifier operated such that ½V_{sat}½ = 10V. Calculate the RC time-constant required to produce a 5 ms output pulse.

Solution

A) First calculate β:

$$\beta = \frac{R3}{R2 + R3} \qquad \text{eq. (7-24)}$$

$$\beta = \frac{3.3\ kohms}{10\ kohms + 3.3\ kohms} \qquad eq.\ (7\text{-}25)$$

$$\beta = \frac{3.3\ kohms}{13.3\ kohms} = 0.248 \qquad eq.\ (7\text{-}26)$$

B) Calculate the RC time-constant:

$$RC = \frac{T}{\text{LN}\left[\dfrac{1 + 0.7\,V / V_{sat}}{1 - \beta}\right]} \qquad eq.\ (7\text{-}27)$$

$$RC = \frac{\left[5\ ms\ \_\ \dfrac{1\ \text{sec}}{1000\ ms}\right]}{\text{LN}\left[\dfrac{1 + 0.7\,V / 10\ volts}{1 - 0.248}\right]} \qquad eq.\ (7\text{-}28)$$

$$RC = \frac{0.005\ \text{sec}}{\text{LN}\ [1.07 / 0.752]} = 0.0142\ \text{sec} \qquad eq.\ (7\text{-}29)$$

Refractory Period

At time T2 the output signal voltage V_o switches from $-V_{sat}$ to $+V_{sat}$. Although the output has timed out, the MMV is not yet ready to accept another trigger pulse. The refractory state between T2 and T3 is characterized by the output being in the stable state, but the input is unable to accept a new trigger input stimulus. The refractory period must await the discharge of C1 under the influence of the output voltage to satisfy V1 < (V1 - 0.7) volts.

Chapter 8

Retriggerable Timers

A monostable multivibrator (aka one-shot) is a binary circuit with only one stable state. There are two *permissable* states (HIGH and LOW), but only one of these is stable. The monostable multivibrator is designed to remain in whichever of the two states is the dormant condition until a trigger pulse causes it to shift to the transient state. But the transient state is unstable. The circuit will remain in that state only for a specified, predetermined period of time *T*. After that period expires, the circuit reverts to the dormant state. It will then remain dormant until another trigger pulse is received. In most circuits, the duration of the transient time is set by an RC time-constant.

Many multivibrator circuits also have a brief refractory period following the transient or active period, in which no further trigger pulses are honored. This is usually the period required for the capacitor in the RC timing network to recharge (or discharge in some cases) back to its dormant condition.

Regardless of whether or not there is a refractory period, most monostable circuits cannot be retriggered until the transient period, *T*, has expired. Those that have refractory periods require the transient period plus the refractory period before any additional trigger pulses will be honored. This behavior is not a circuit defect; it is absolutely essential in some applications. For bounceless pushbuttons, for example, the whole purpose of the one-shot is defeated if the circuit responds to additional trigger pulses. In that application, the first bounce of the switch's mechanical contacts will trigger the one-shot, which proceeds to trigger some additional circuitry while ignoring the multiple secondary bounces of the switch contacts.

But there are applications where it would be nice to retrigger a one-shot. Such applications include any circuit where the one-shot is triggered every time some event occurs. If the event occurs again before the circuit is ready to accept another trigger pulse, then some function of the circuit will be lost. In those cases, a *retriggerable multivibrator* is needed.

Consider the example in **Figure 8-1**, which is the block diagram for a medical respirator alarm. Respirators have a mode called Intermittant Mandatory Ventilation (IMV). The IMV mode is used to wean the patient off the respirator by reducing the frequency of breaths from the respirator gradually until it reaches one per minute or less. If the patient ceases breathing on his or her own, then it could be almost 60 seconds before the respirator pulses them. The need is for an alarm that will detect whether or not the patient is breathing.

A sensor is placed in the exhalation line so that it provides a signal regardless of whether the patient breaths on his/her own or the respirator provides the breathing. The sensor used in one alarm was a pair of thermistors in a bridge circuit. When operated close to the point of self-heating, small breath variations cause a waveform such as shown in **Figure 8-1**.

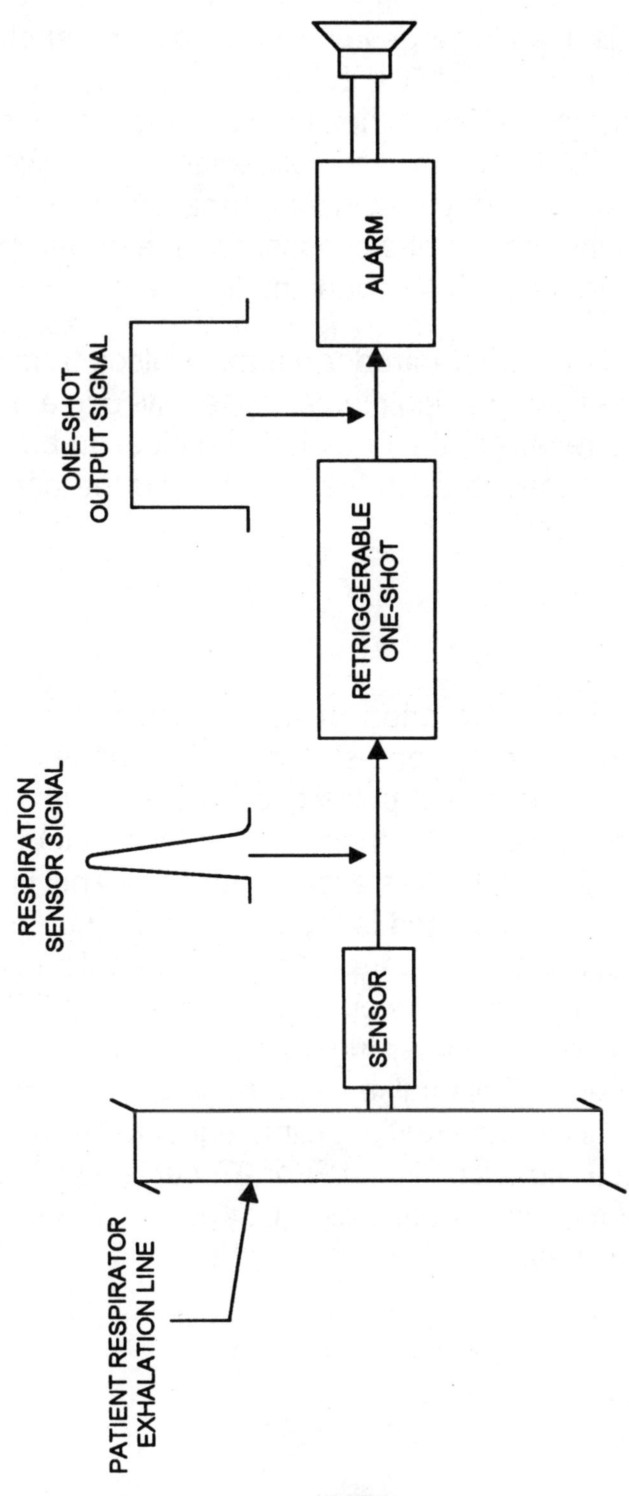

Figure 8-1. Application of a retriggerable one-shot in a medical alarm circuit.

The breath signal is used to trigger a retriggerable one-shot. As long as the output of the one-shot is HIGH, the alarm does not sound. But if the one-shot output drops LOW then the alarm sounds. The duration of the one-shot is set to a bit longer than the expected time between successive breaths, with a little extra margin for normal variation. If the patient ceases breathing, the retriggering ceases, and the one-shot times-out, causing its output to drop LOW...sounding the alarm.

A similar concept is used for intruder alarms. Pulses from an infrared light source, e.g., a light emitting diode, are received in a photocell. As long as pulses are received, the one-shot continues to be retriggered. But if an intruder passes through the beam, then the one-shot times out and the alarm sounds.

Operation

Figure 8-2 shows the operation of the retriggerable monostable multivibrator. The period of the one-shot pulse is *T*. At time T1, a trigger pulse is received, so the output pulse goes HIGH for time *T*. No additional pulses are received on the trigger line, so the output drops LOW again as soon as its period *T* has expired (i.e., T2). This is normal operation for a one-shot circuit and is the same as for non-retriggerable circuits. At time T3, another trigger pulse is received. The output of the one-shot again goes HIGH and would remain HIGH until *T* expires (at point T5). But at time T4 another trigger pulse is received prior to the expiration of the period. This pulse causes the circuit to retrigger; the output remains HIGH for an additional time equal to period *T*. The total duration of the HIGH condition is not *T*, or 2T, but is equal to *T* plus the expired portion of the first output pulse (i.e., duration T4-T3). The total output pulse duration is, therefore:

$$T' = T + (T4 - T3) \qquad \text{eq. (8-1)}$$

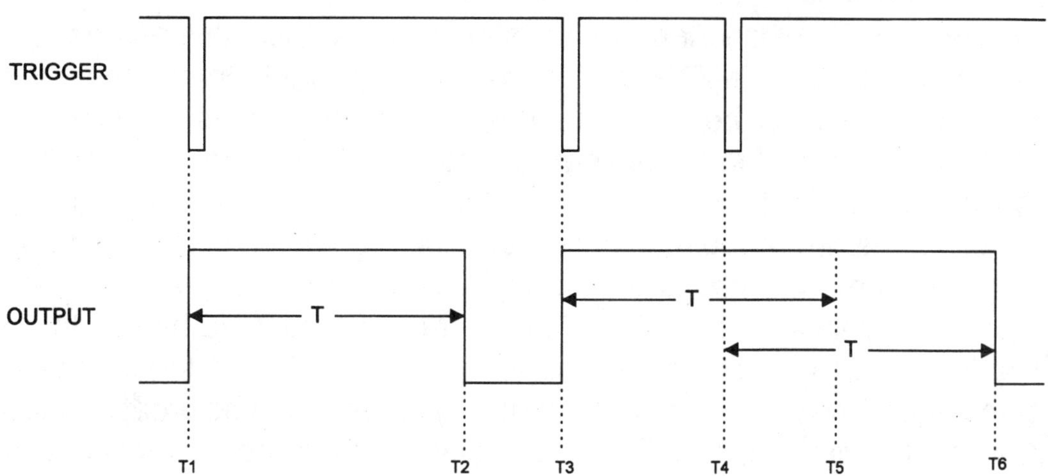

Figure 8-2. Waveforms of non-retriggered and retriggered operation.

This is the behavior expected of a retriggerable monostable multivibrator. There are several different types of retriggerable circuits available. Some are special purpose chips (i.e., 74122), while others are clever circuits involving normally non-retriggerable circuits.

The LM-555 timer IC can be operated as a non-retriggerable monostable multivibrator circuit, with an output duration that is a function of the RC time-constant R1C1.

Figure 8-3a shows a method for making the 555 retriggerable from a negative-going pulse. In ordinary operation, the trigger input of the 555 (pin no. 2) is held HIGH (i.e., greater than 0.333(V+)), usually very close to the V+ voltage. A negative-going trigger pulse must pull the trigger input LOW to a voltage less than 0.333(V+). If this is done, then an internal comparator sets an RS flip-flop and the output of this flip-flop is inverted to form the chip output.

The circuit is not retriggerable because the charge on the capacitor is not dumped by the trigger pulse. As long as the capacitor continues to charge, and is less than 0.667(V+), then the output remains HIGH and the trigger input is not affected. The trick is to reset the internal RS flip-flop by discharging capacitor C1 using the trigger pulse.

In **Figure 8-3a** we connect a transistor switch across timing capacitor C1. Transistor Q1 is a PNP type of device, so it wants to see the base more negative than the emitter in order to be forward-biased. The emitter of Q1 is connected to the ungrounded end of capacitor C1, so Q1 is forward-biased if the trigger line is HIGH. The negative-going trigger pulse serves two purposes: it will cause the trigger line of the 555 to be activated (i.e., brought to a potential less than 0.333V+), and it will momentarily forward-bias Q1. With the transistor forward biased there is a low-impedance short circuit across the capacitor, so the capacitor discharges. The duration of the retriggering pulse must be great enough to allow the discharge of C1 to take place. This is not too much of a problem in most cases, unless the amplitude of the trigger pulse is insufficient to drive the transistor into saturation.

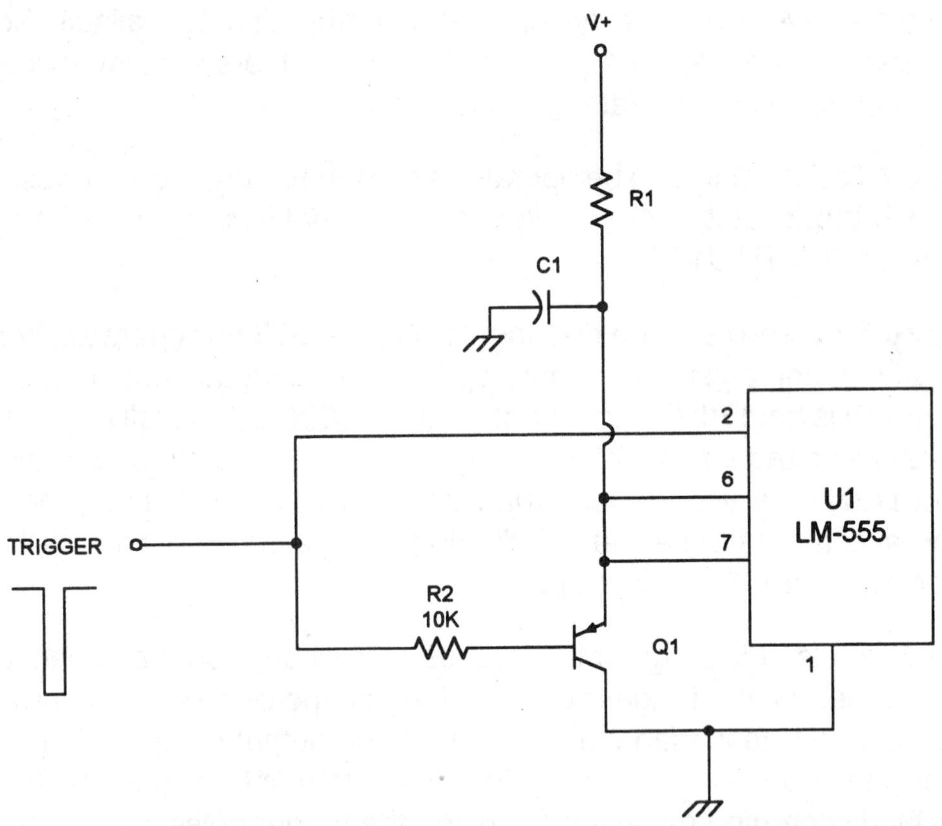

Figure 8-3a. LM-555 retriggerable circuit responds to negative-going trigger pulses.

Another method is shown in **Figure 8-3b**, and this method allows the use of a positive-going trigger pulse. Two transistors are used: one is connected across C1 and will cause it to be discharged whenever the transistor is forward-biased. The other transistor is connected to the trigger terminal of the LM-555. The collector of transistor Q2 is also connected to V+ thropugh a pull-up resistor. This arrangement ensures that the trigger terminal of the 555 is held HIGH when no trigger pulse is present. But when a positive-going trigger pulse is applied to the bases of the two transistors, the 555 will be retriggered. When transistor Q1 is forward-biased, the charge (if any) in capacitor C1 is dumped. Also, when transistor Q2 is forward-biased it grounds pin no. 2 of the 555 for the duration of the trigger pulse.

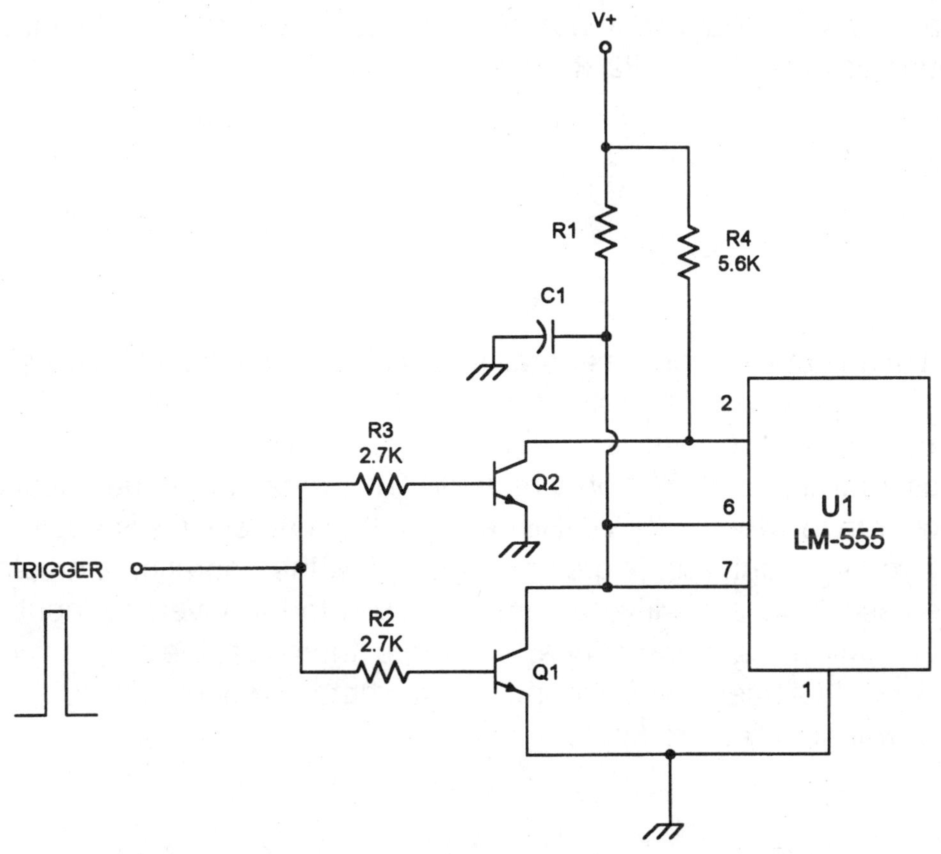

Figure 8-3b. LM-555 retriggerable circuit responds to positive-going trigger pulses.

A similar tactic is used in **Figure 8-4**, this time using the XR-2240 IC timer. This method also works with the 8240 second-source version of the 2240.

The XR-2240 timer requires a positive trigger pulse applied to pin no. 11 of the chip. But we have the same problem on the XR-2240 that exists on the 555. The timer will not honor additional trigger pulses until the capacitor has charged to the threshold level required to time-out the chip. The solution is the same: we add transistor switch Q1 to the circuit in order to dump the charge on capacitor C1 in response to trigger or retrigger pulses.

An example of an operational amplifier retriggerable monostable multivibrator is shown in **Figure 8-5a**. The circuit uses the op-amp in its capacity as a voltage comparator. The noninverting input is biased through voltage divider R2/R3 to a value of:

- $$\beta V = \frac{R3\,(V+)}{R2 + R3} \qquad \text{eq. (8-2)}$$

The timing network consists of resistor R1 and capacitor C1. There is a junction field effect transistor (JFET), Q1, in parallel with C1. In the dormant condition, however, JFET Q1 is pinched off by the V- DC power supply (through the 470 kohm resistor). This means that the capacitor will charge to very nearly V+. Since the voltage across C1 is applied to the inverting input, and is greater than β(V+), the operational amplifier sees essentially a positive potential applied to the inverting input. By the ordinary rules of operational amplifiers, therefore, the output will be LOW- or HIGH-negative, depending on whether monopolar or bipolar DC power supplies are used.

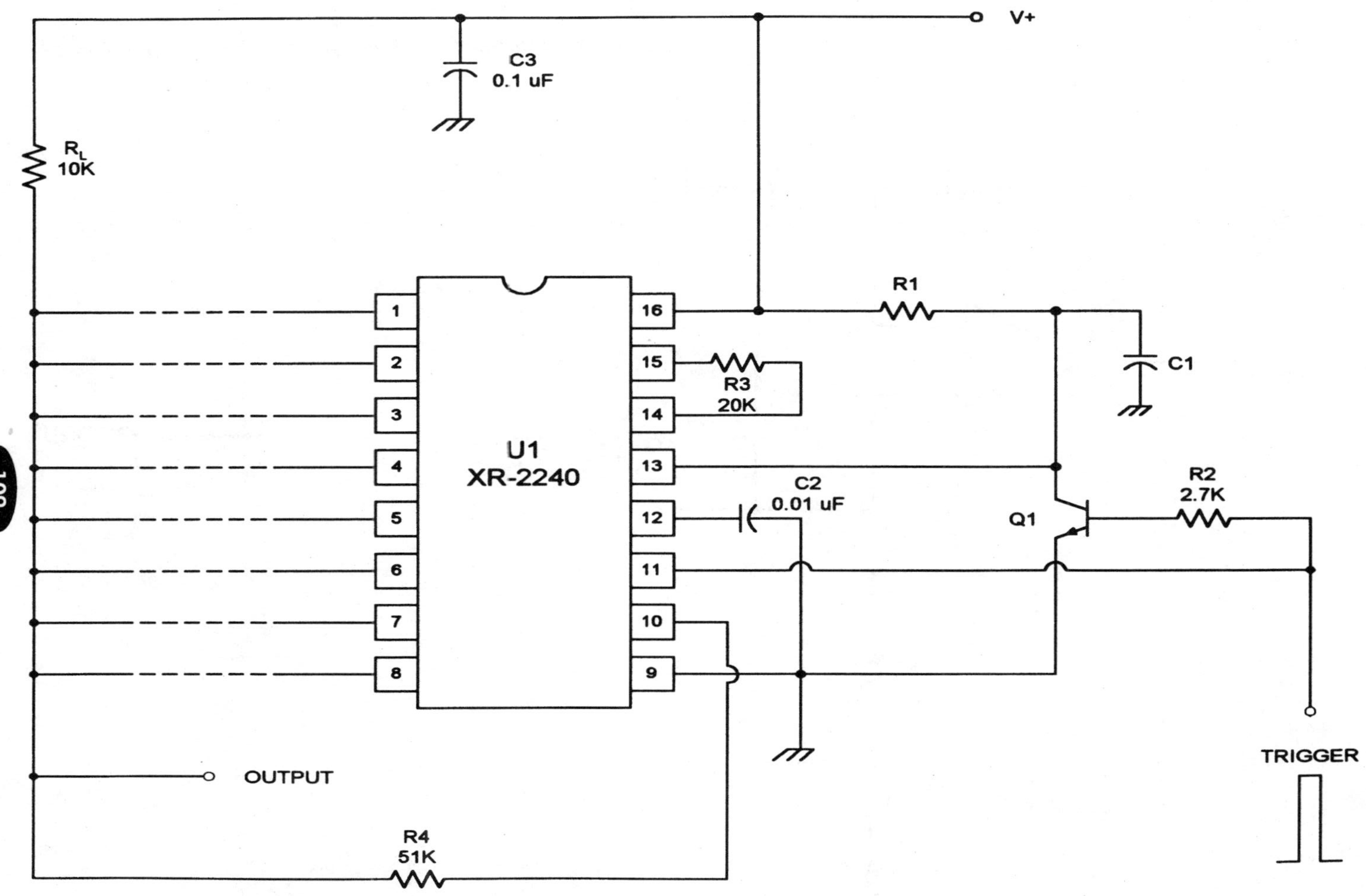

Figure 8-4. XR-2240 retriggerable one-shot circuit. Open R4 for stable operation.

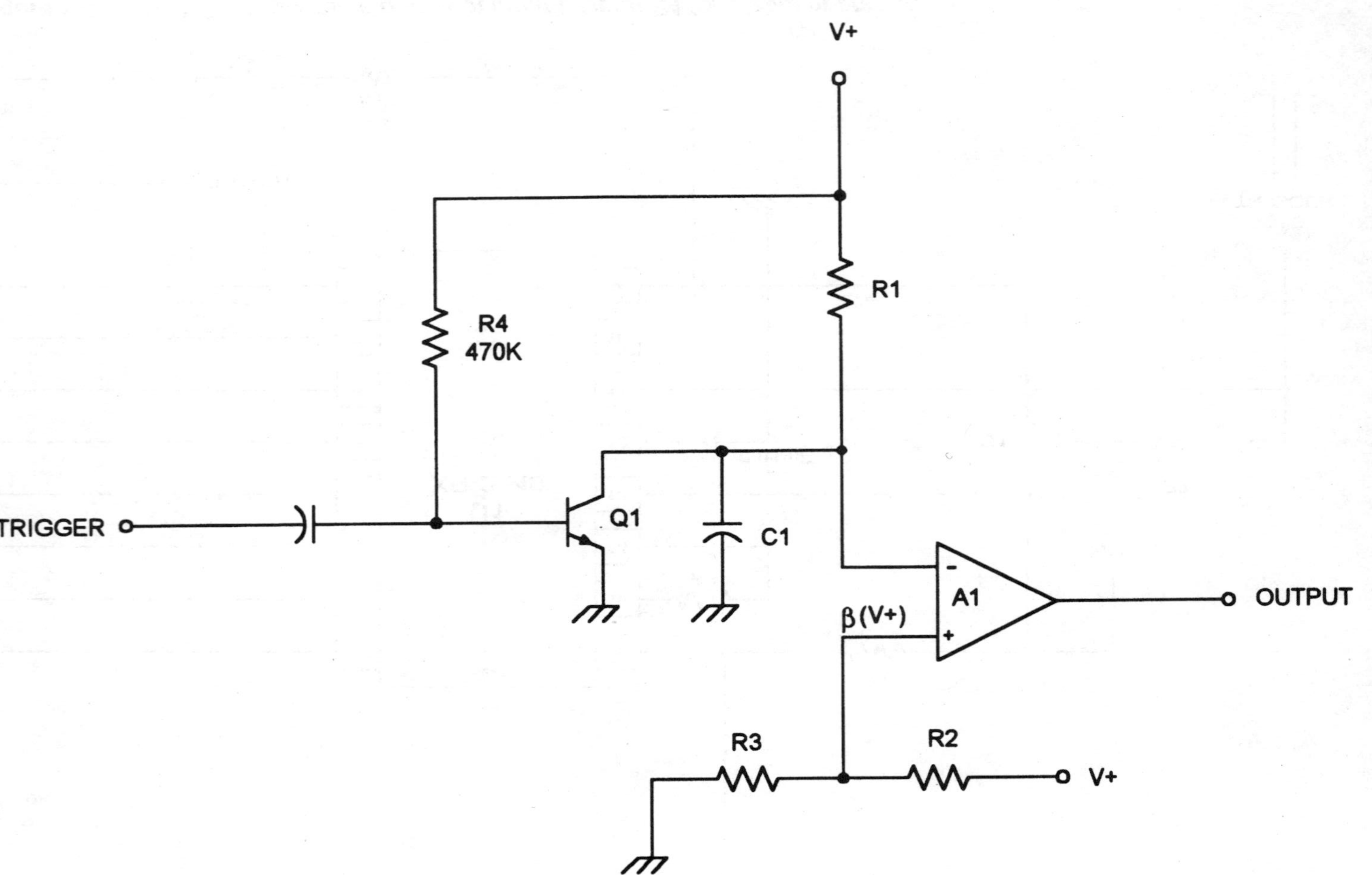

Figure 8-5a. Op-amp retriggerable one-shot circuit.

When a positive-going pulse is applied to the gate of the JFET, then the channel resistance of the JFET goes from very high to very low, thereby discharging the capacitor. The input of the operational amplifier now sees a negative voltage because β(V+) is greater than zero. The op-amp output snaps HIGH-positive, and, after the trigger pulse vanishes, capacitor C1 begins to recharge.

The output of the operational amplifier remains HIGH for the period required to recharge to β(V+). At that time, the operational amplifier output drops LOW again. We can retrigger the circuit at any time because the trigger pulse drops the charge on C1 to zero, and the process has to begin anew.

The timing waveforms for this circuit are shown in **Figure 8-5b**. The normal monostable operation is shown at time T1. When the positive-going trigger pulse is received, it rapidly discharges C1 from V+ to zero. As soon as the voltage across C1 drops lower than β(V+), the output of the op-amp snaps HIGH. When the capacitor voltage rises to β(V+), then the comparator sees a zero input condition, so the output voltage drops LOW.

The retriggerable operation of the circuit is shown at T3 to T6. A trigger pulse is received at time T3, and the same operation takes place. But, at time T5, another trigger pulse is received. This pulse turns on the JFET, thereby forcing the charge on capacitor C1 back to zero. The output state does not change because the op-amp input is relatively the same—at least in polarity.

The duration of the output pulse is given by the expression:

- $$T = R1\,C1\,Ln\left[\frac{R3}{R2} + 1\right] \qquad \textit{eq. (8-3)}$$

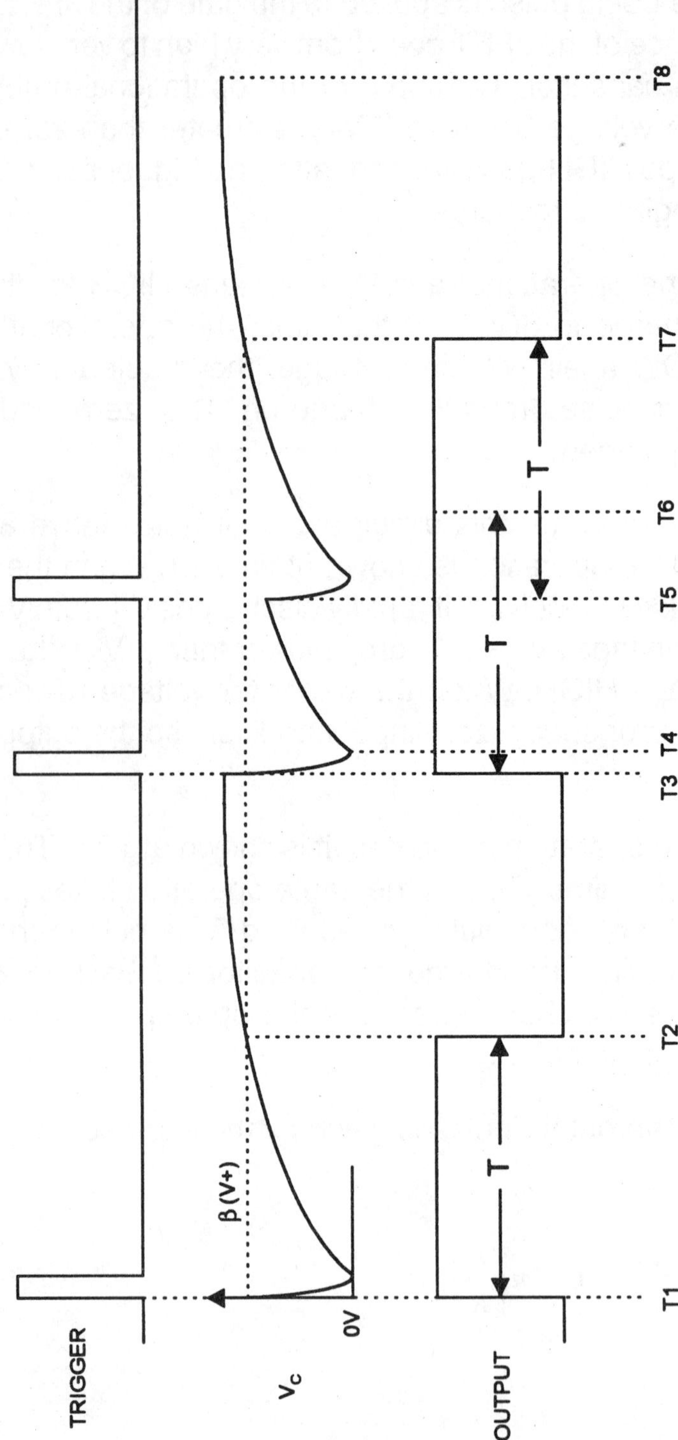

Figure 8-5b. Timing waveforms of the op-amp retriggerable one-shot circuit.

The total time that the monostable will be in the output HIGH state in the example in **Figure 8-5b** is:

$$T' = T + (T5 - T3) \qquad \text{eq. (8-4)}$$

TTL Retriggerable One-Shots

In Chapter 2 you were introduced to TTL monostable multivibrator chips, one of which was the 74122 device. A general circuit for these circuits is shown in **Figure 8-6**. This particular version uses the 74122 device, which is a single retriggerable monostable multivibrator from the TTL family. The device has complementary outputs, so it will produce both Q and NOT-Q levels. The trigger is a negative-going pulse that makes the transition from +5 volts to zero when triggering the one-shot. Resistor R1 and capacitor C1 form the timing for the one-shot, and obey the expression:

$$T = 0.37\, R1\, C1\, Ln\left[\frac{0.7}{R1} + 1\right] \qquad \text{eq. (8-5)}$$

Where:

R1 is the resistance of R1 in kohms

C1 is the capacitance of C1 in picofarads

T is the time in nanoseconds (ns)

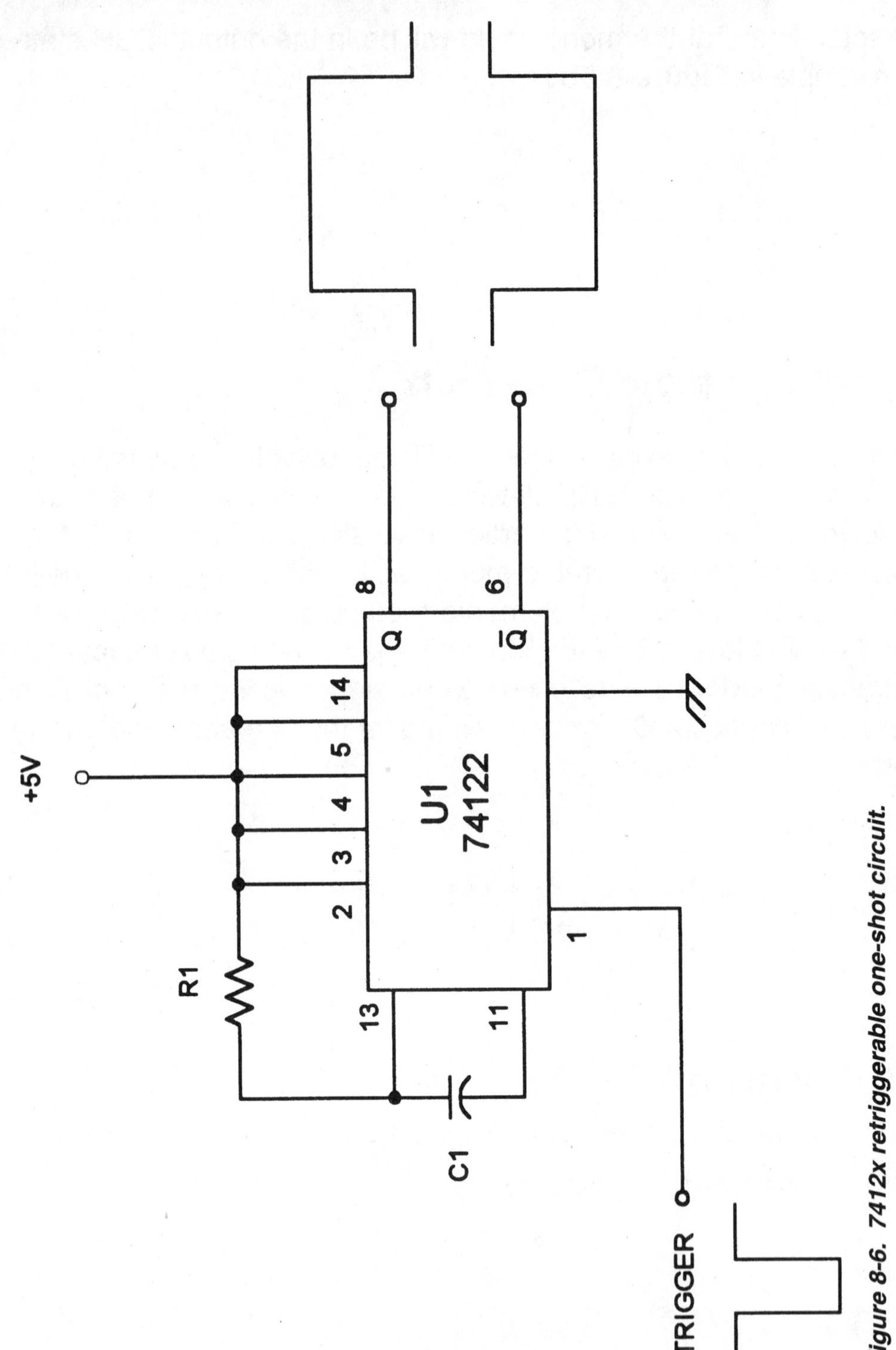

Figure 8-6. 7412x retriggerable one-shot circuit.

A certain minimum time, a refractory period of sorts, must expire before the 74122 can be retriggered:

$$T_{RT} = 0.22\ C1 \qquad \textit{eq. (8-6)}$$

Where:

T_{RT} is the time to retriggerability in nanoseconds

C1 is the capacitance of C1 in picofarads

The 7412x series of monostable multivibrator chips are usually not the best choice in practical circuits unless very fast pulses are needed. If you have have a pulse in, say, the millisecond range, then it would be better to use the 555, XR-2240 or an operational amplifier one-shot circuit. The 7412x are more susceptible to noise triggering than the other devices. The CMOS 4528 device is also a good choice, and is a retriggerable one-shot that does not need extra components to effect retriggering.

An Op-Amp Retriggerable Monostable Multivibrator

The one-shot circuits of Chapter 7 are non-retriggerable MMVs. Once they are triggered the circuit will not respond to further trigger inputs until after both the quasistable and refractory states are completed. A *retriggerable monostable multivibrator* (RMMV) will respond to further trigger signals.

Figure 8-7 shows the RMMV response. An initial trigger signal (V_t) is received at time T1. The output snaps LOW and, under normal circumstances, it would remain in this quasistable state until time T3 when the duration, *T*, expires. But at time T2 a second trigger pulse is received. The circuit is now retriggered for another duration, *T*, so will not time-out until T4. The total time that the RMMV is in the quasistable state is [T + (T2-T1)]. In other words, the RMMV output is active for the entire duration, *T*, plus that portion of the previous active time which expired when the next trigger pulse was received.

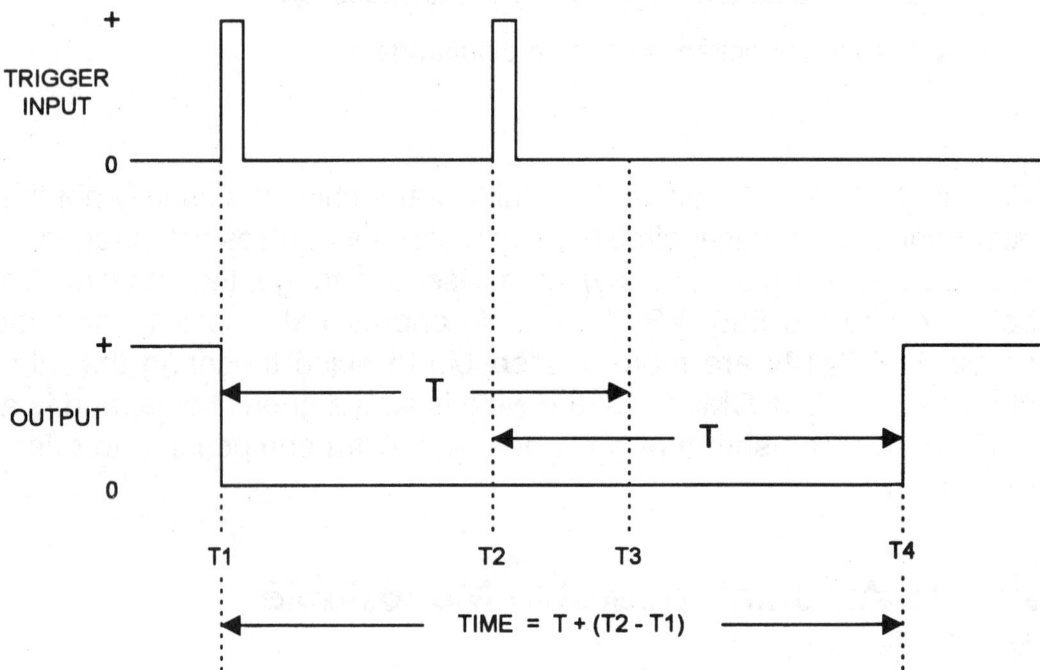

Figure 8-7. Retriggerable operation.

Figure 8-8a shows the circuit for a simple RMMV based on an operational amplifier. The two inputs are biased from a reference voltage source, $+V_{ref}$. The potential applied to +IN is a fraction of $+V_{ref}$. That is, $[(R3)(+V_{ref})/(R2 + R3)]$. The potential applied to -IN is a function of $+V_{ref}$ and time-constant R1C1. If the circuit is not triggered at turn-on, then

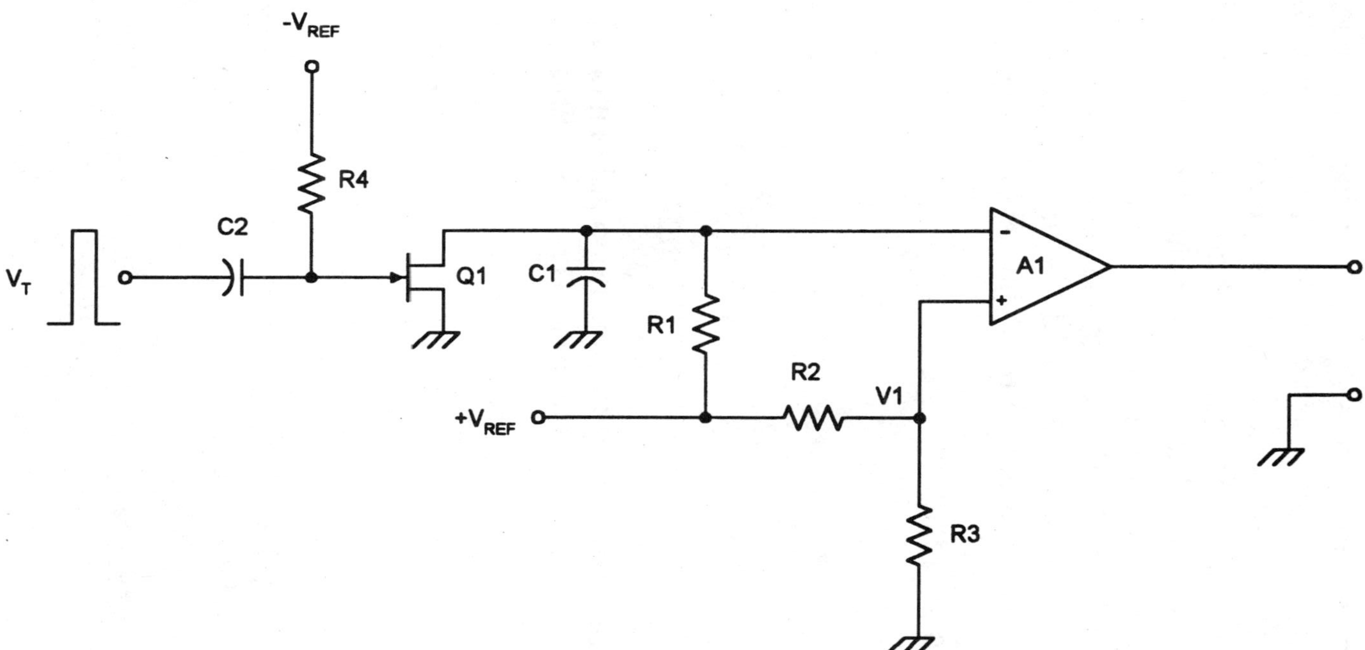

Figure 8-8a. Retriggerable monostable multivibrator based on a single operational amplifier.

the capacitor (C1) charges up to $+V_{ref}$, so -IN is more positive than +IN. This situation forces V_o to $-V_{sat}$, which is the stable state. When a positive-going trigger pulse (V_t) is received (see **Figure 8-8b**), it biases the JFET, Q1, hard on. The JFET drain-source channel resistance drops to a very low value, causing C1 to discharge rapidly between T1 and T2. With V_c close to 0 VDC, +IN is more positive than -IN, so the output snaps abruptly to $+V_{sat}$ at time T1. During the interval T2 to T3 capacitor C1 begins charging towards $+V_{ref}$, and V_o remains at $+V_{sat}$. Once V_c reaches +V1, however, the output of A1 snaps back to $-V_{sat}$.

The duration, *T*, is found from:

$$T = R1C1 \ \mathrm{LN}\left[\frac{R3}{R2} + 1\right] \qquad \text{eq. (8-7)}$$

Example

Calculate the duration of a retriggerable monostable multivibrator (**Figure 8-8a**) in which R2 = R3 = 10 kohms, R1 = 15 kohms and C1 = 0.1 µF.

Solution

$$T = R1C1 \ \mathrm{LN}\left[\frac{R3}{R2} + 1\right] \qquad \text{eq.(8-8)}$$

$$T = 15\,k\Omega \times \left[0.01\,\mu F \times \frac{1\,F}{10^6\,\mu F}\right] \times \mathrm{LN}\left[\frac{10\,k\Omega}{10\,k\Omega} + 1\right]$$

eq.(8-9)

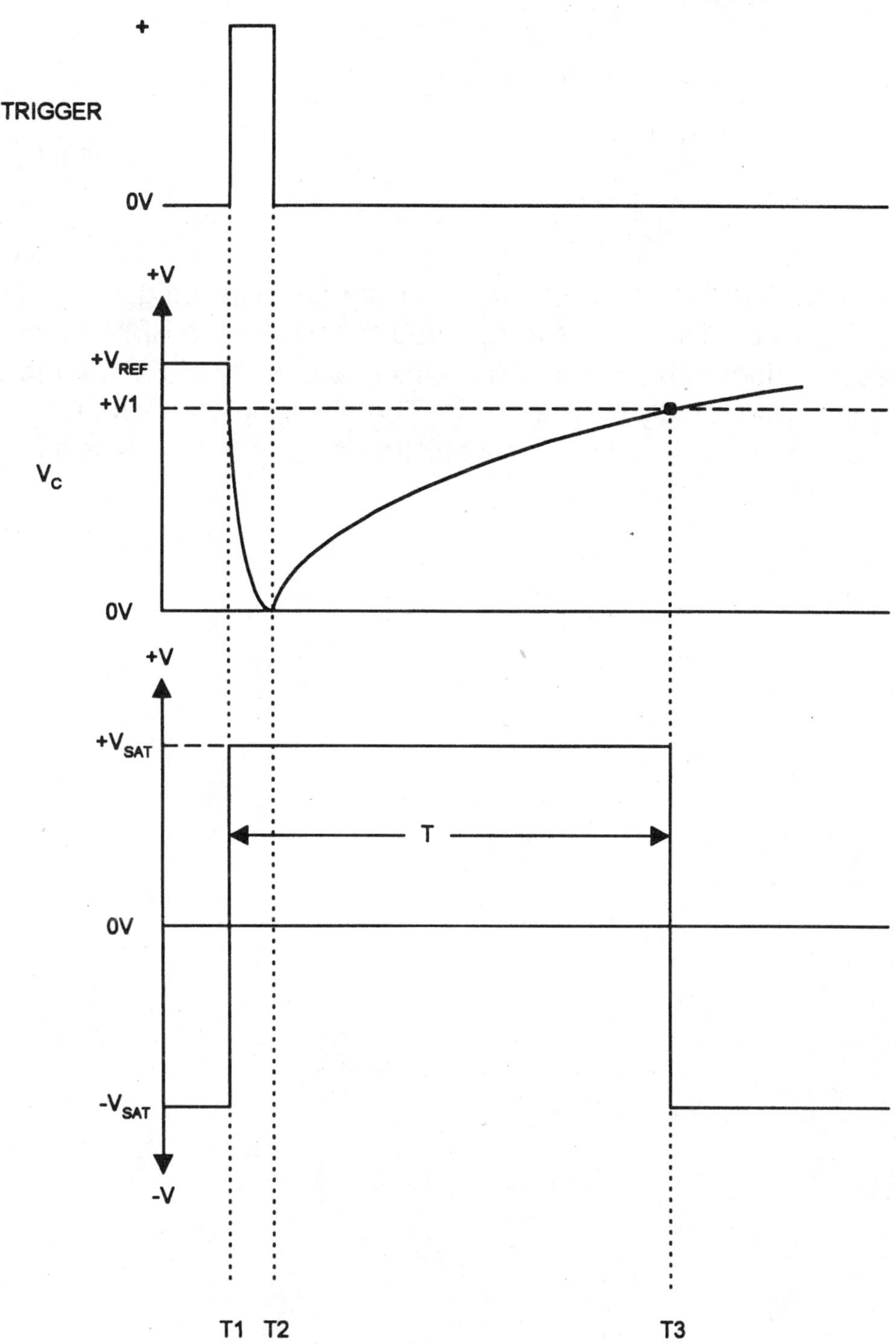

Figure 8-8b. Nonretriggered operation.

- $T = (1.5 \times 10^{-4}) \times \mathrm{LN}\ (2)\ \mathrm{sec} = 1.04 \times 10^{-4}\ \mathrm{sec}$

eq.(8-10)

The operation discussed above is for normal non-retriggered operation. **Figure 8-8c** shows the retriggered case. Here the RMMV receives a second trigger pulse at time T2, which forces the JFET (Q1) to turn on again, and rapidly discharge C1. The charging process then starts over again and will continue until the circuit times out...unless a further trigger pulse is received.

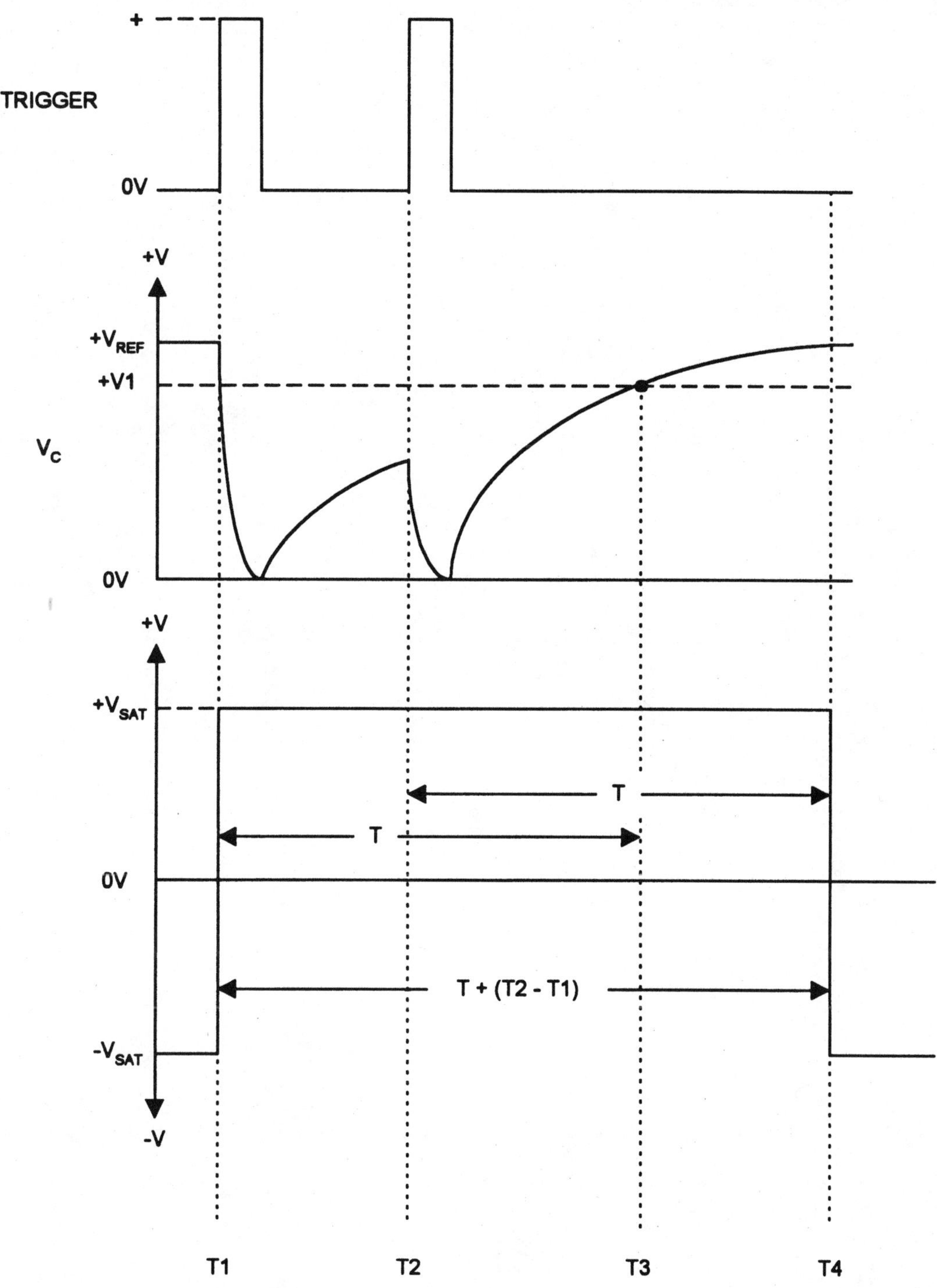

Figure 8-8c. Retriggered operation.

Chapter 9

Long Duration Timers

The long duration timer presents a special problem to the electronic circuit designer. The drift and other errors of some components tend to accumulate, and the total error over long time periods tend to be large. In fact, whenever we see errors that are functions of time, then the long duration timer will suffer markedly. Any long duration RC timer will, for example, suffer from several time-related problems. We must find some way to overcome these errors if we are to use an electronic circuit for long durations.

A viable alternative, incidentally, is to use electromechanical timer elements in long-duration application. We can buy timer elements that can be either manually or electrically set to whatever duration we want, up to about 48 hours. The time will either close or open (depending on design) an electrical switch when the set duration has expired. But this is not an elegant approach and is not satisfying to the electronic designer. How do the various circuits perform? Our example will be the XR-2240 RC timer chip because it has attributes that make it ideal for long duration circuits.

The timebase section of the XR-2240 IC timer suffers from the same problem as the LM-555 device. For long duration timing we need electrolytic capacitors because they are high value. Unfortunately, some of those capacitors have a capacitance that is between -20% and +100% of the marked value...and it tends to drift with temperature. In addition, some forms of electrolytic capacitor have high values of shunt resistance. In short duration circuits the shunting effect is negligible, but when the R1 timing resistor is also high the total resistance changes appreciably.

The advantage of the XR-2240 device is that it contains an internal binary counter (see Chapter 6), and it can be used to form very long duration timers. We are then able to use lower-valued components that have tigher tolerances and lower temperature coefficients.

Recall from Chapter 6 that the duration of the output pulse, when the wired-OR output configuration is used, can be anything from 1RC to 255RC. If we set the RC time-constant to produce a basic duration of 1 second, then we can make a 1-second to 255-second timer, depending on which outputs are wired together, but which has the accuracy and temperature coefficient characteristics of the 1-second timer. We can use time-constants as long as 10 seconds safely in the XR-2240 circuit. This means that a single XR-2240 circuit can produce timers up to 2,550 seconds (i.e., 42.5 minutes). Greater times than that, however, require that we either use some other device or cascade two or more XR-2240 devices.

Figure 9-1 shows two XR-2240 timers in cascade to produce a very long duration timer with the stability and accuracy of a short duration circuit. The first XR-2240 is used as the timebase for the entire circuit, and has a frequency set by R1 and C1 (T = R1C1). In this case, we use

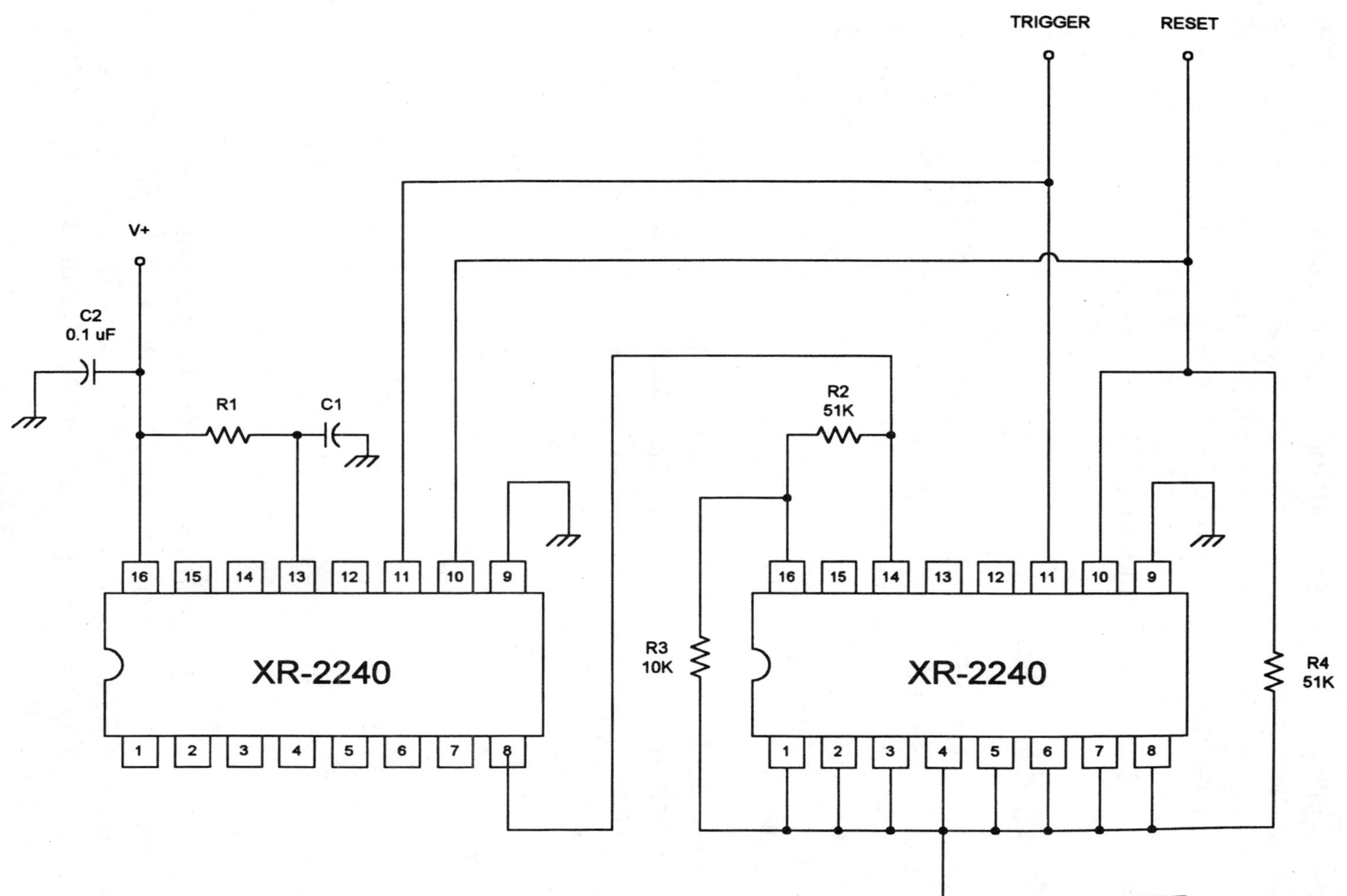

Figure 9-1. Long duration timer using two XR-2240 devices.

only the eighth bit of the binary counter. The output of the first XR-2240 is used as the timebase pulse for the second XR-2240. All of the outputs of the second unit are wired together (wired-OR configuration), so the output time is $256^2 \times T$, or $65{,}536 \times T$, where *T* is the period of the RC time-constant. The output period T_O, when the RC time-constant *T* is one second, therfore, would be 65,536 seconds, or 18.2 hours.

XR-2240 Examples

The circuit shown in **Figure 9-2** is a 10-second timer. Originally, this timer circuit was used by amateur radio operators to warn them of the 10-second deadline for transmitting callsign identification. A period of 10 minutes is 600 seconds, and this circuit times-out in 572 seconds ...just in time for the operator to make the required transmission of his or her callsign.

All of the outputs of the XR-2240 are wired together in the wired-OR configuration. This means that the output duration will be 255T, where T = R1C1. The value of *T* in this case is $(3.3 \times 10^6\ \Omega) \times (0.68 \times 10^{-6}$ farads$)$ = 2.244 seconds, so the output will be:

- $$T = (2.244\ \text{s})(255) \qquad \textit{eq. (9-1)}$$

- $$T = 572.2\ \text{seconds} \qquad \textit{eq. (9-2)}$$

The circuit is a one-shot multivibrator. Applying a positive level to the trigger input (pin no. 11) by pressing the switch (S1) will initiate the timer sequence. The output will drop LOW for 572 seconds. When it goes HIGH again a signal is applied to pin no. 10 that resets the counter to its dormant state.

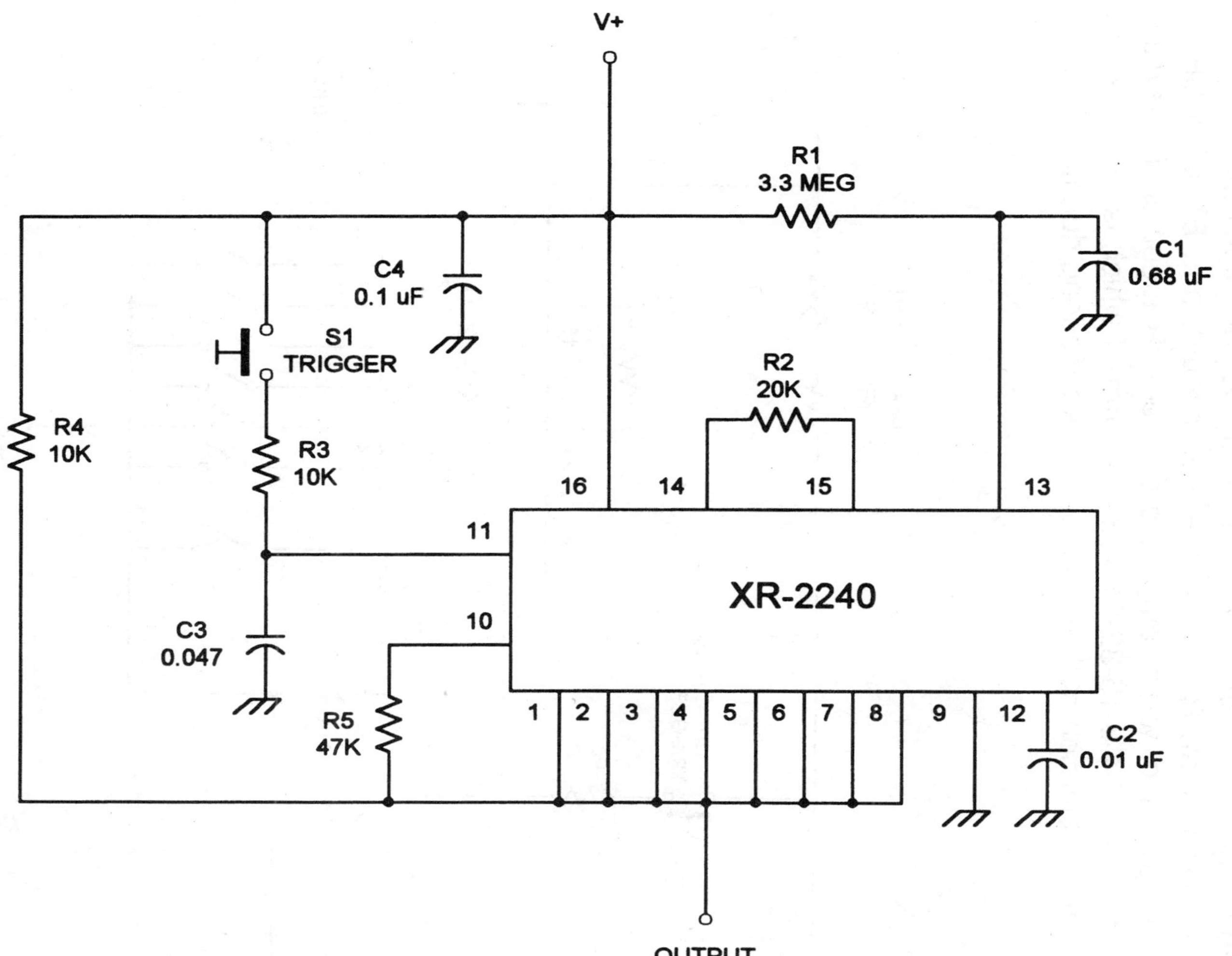

Figure 9-2. Ten-second timer.

A variable output timer circuit is presented in **Figure 9-3**. It is similar in concept to **Figure 9-2** except that the timing resistor is made variable by: A) using a fixed resistor in series with a potentiometer (R1A and R1B in series form R1 in the previous circuit); and B) using SPST switches to select which outputs are wired-OR together. The switch sequence is 1-2-4-8-16-32-64-128, the sum of which is 255. We can use any combination of these settings to form a long duration timer.

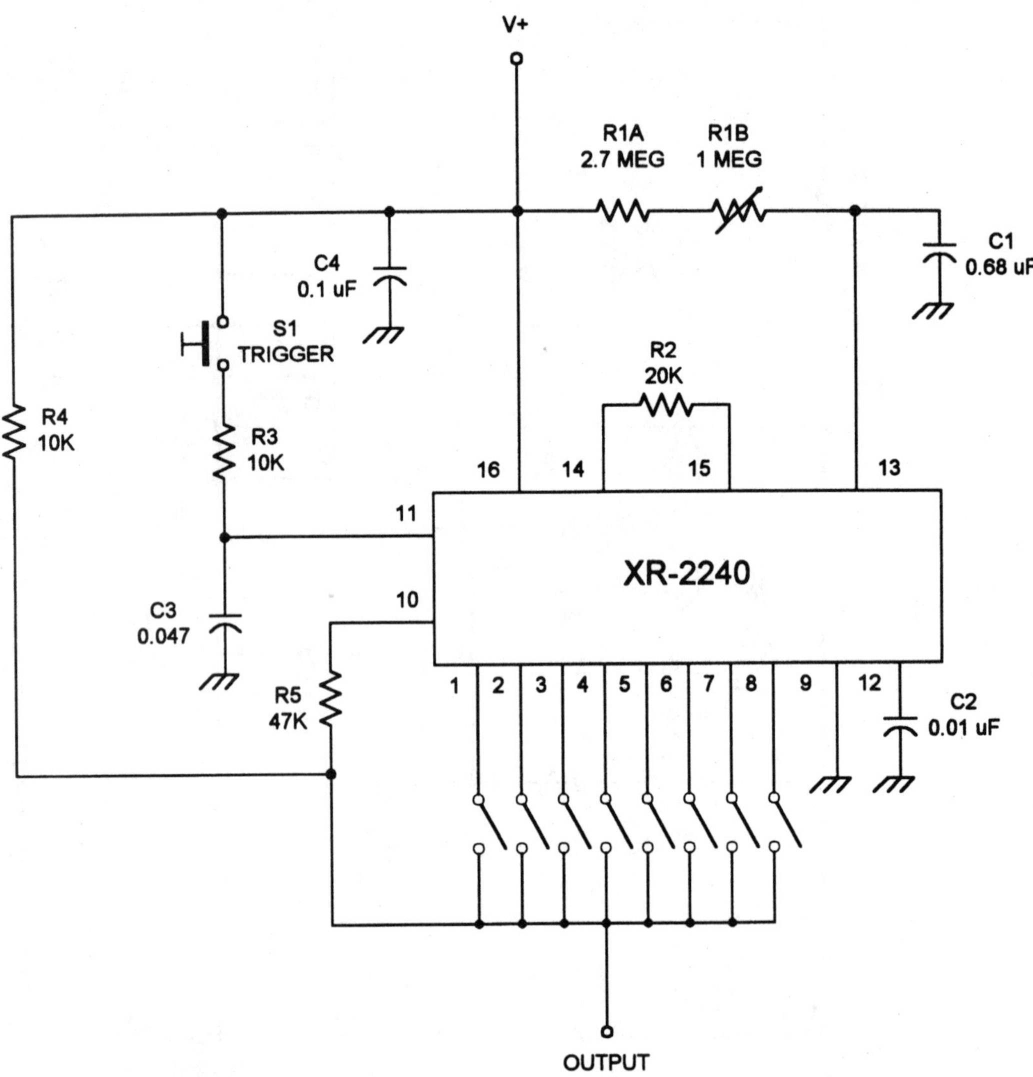

Figure 9-3. Variable long duration timer.

CMOS Example

The circuit in **Figure 9-1** is probably the cheapest and most viable method for obtaining very long duration timers. However, we can also press certain CMOS devices into service as well. In **Figure 9-4** we see the use of the CMOS 4060 device as a long duration timer. The 4020 device will also work. This timer is based on the 4060, which is a fourteen-stage ripple counter, so it can output pulses with division ratios over the input frequency up to 2^{14} or 16,384. If we apply a 0.5 Hz clock signal to the input of the IC CMOS counter, then the 2^{14} output will go HIGH once every 2 sec. $\times$ 2^{14} = 32,768 seconds. This period is 9.1 hours. Since we can use clocks to about 0.1 Hz (10-second clock pe-

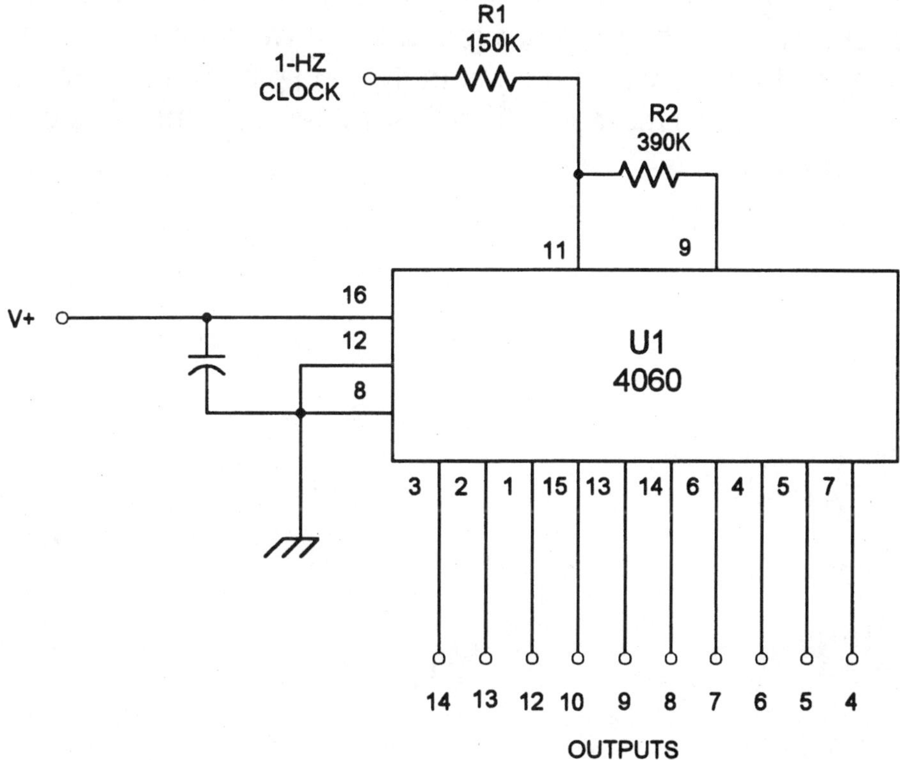

Figure 9-4. CMOS long duration timer.

riod) by using the correct RC time-constant in the clock, we can build timers to 163,840 seconds, or 45.5 hours. These circuits incorporate relatively simple and easy-to-use R and C components.

The 4060 contains its own clock, but in this circuit we are not using the internal clock. The clock terminals are wired together to form a Schmitt trigger. This connection means that we must supply a noise-free clock pulse with a good strong amplitude. There are not outputs for the 2, 4, 8 or 2,048 counts, but we can select any other base-2 output up to 16,384. Using the outputs directly limits us to durations that are powers of two, but using them in combination, with gate circuitry external to the 4060, allows us to select any period that is an integer multiple of the input time period.

The reset or RST terminal, pin no. 12, is normally kept LOW for continuous operation, so is shown here grounded. If we want to reset all outputs to zero, then we would momentarily bring pin no. 12 HIGH. But for the type of operation represented in **Figure 9-4** we want to keep pin no. 12 grounded.

Timer Projects and Circuits

Touchplate Trigger
Missing Pulse Detector
Pulse Position Circuit
Tachometry and One-Shot Circuits
Analog Audio Frequency Meter
One-Second Timer/Flasher
Level Translation of LM-555 Output
100-kHz Oscillator
Relay and Optoisolator Drivers
Frequency Synchronized Operation
Code Practice Oscillator
Tone Generator
Warble Siren/Alarm
Two-Phase Digital Clock
Sixteen-Phase Clock
Power Failure Alarm
DC-to-DC Converter
Visual Continuity Tester
Aural Continuity Testers
100-kHz Crystal Calibrator
Pulse Catcher Circuit

Chapter 10

Touchplate Trigger

A touchplate trigger circuit is shown in **Figure 10-1a**. It will trigger a one-shot LM-555 circuit when it is touched, i.e., when finger resistance appears across the electrode ends. The pull-up resistor R2 has a very high value (22 megohms shown here). The touchplate consists of a pair of closely spaced electrodes. As long as there is no external resistance between the two halves of the touchplate, the trigger input of the 555 remains at V+. But when a resistance is connected across the touchplate, the voltage (V1) drops to a very low value. If the average finger resistance is about 20 kohms, the voltage drops to:

$$V1 = \frac{(V+)(20\,k\Omega)}{(R2 + 20\,k\Omega)} \qquad \text{eq. (10-1)}$$

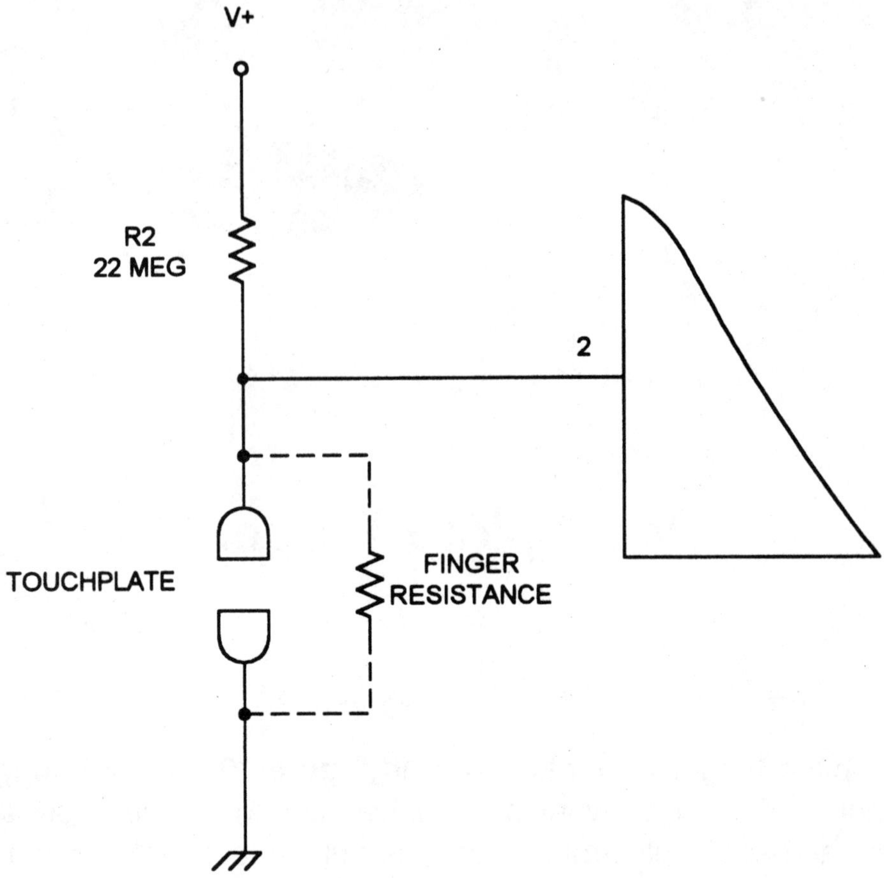

Figure 10-1a. Touchplate circuit.

Which, when R2 = 22 megohms, is equal to 0.0009(V+)— is certainly less than (V+)/3.

The same concept is used in the liquid level detector shown in **Figure 10-1b**. Once again a 22 megohm pull-up resistor is used to keep pin no. 2 at V+ under normal operation. When the liquid level rises sufficiently to short out the electrodes, however, the voltage on pin no. 2 (V1) drops to a very low level, forcing the 555 to trigger.

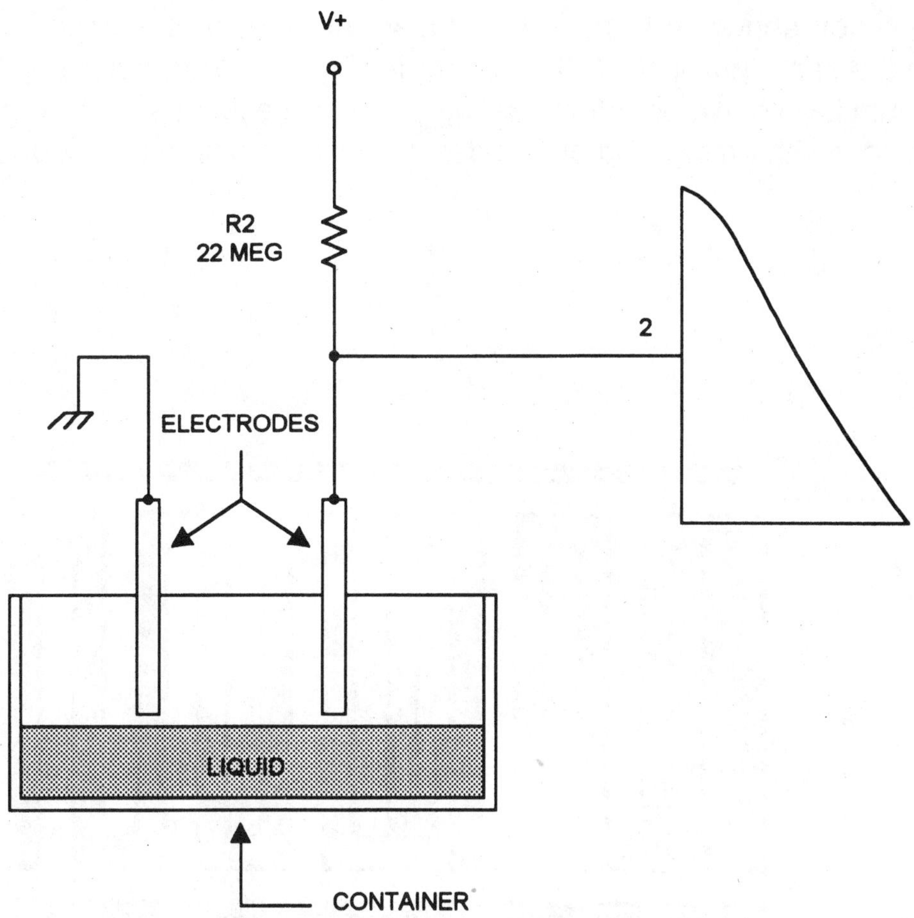

Figure 10-1b. Liquid level detector.

There are two different approaches to sensor design. The version shown in **Figure 10-1c** is made from printed circuit board, or some other means of laying conductors on a flat surface. This interdigital sensor places two sets of printed electrodes interspaced with each other. It can be used for rain sensing. If rain water causes any two adjacent fingers to see a lowered resistance, the trigger occurs.

One problem with this form of sensor is that it must be cleaned occasionally. Minerals in the water will tend to form permanent short circuits. Also, a few small holes should be placed in the insulating substrate to allow water to drain out.

The version shown in **Figure 10-1d** is a cup sensor. It is intended to serve the same purpose as the sensor in **Figure 10-1c**, but requires a minimum level of water before it is triggered. It is essentially like **Figure 10-1b**, but with embedded electrodes rather than dipping electrodes.

Figure 10-1c. Moisture detector.

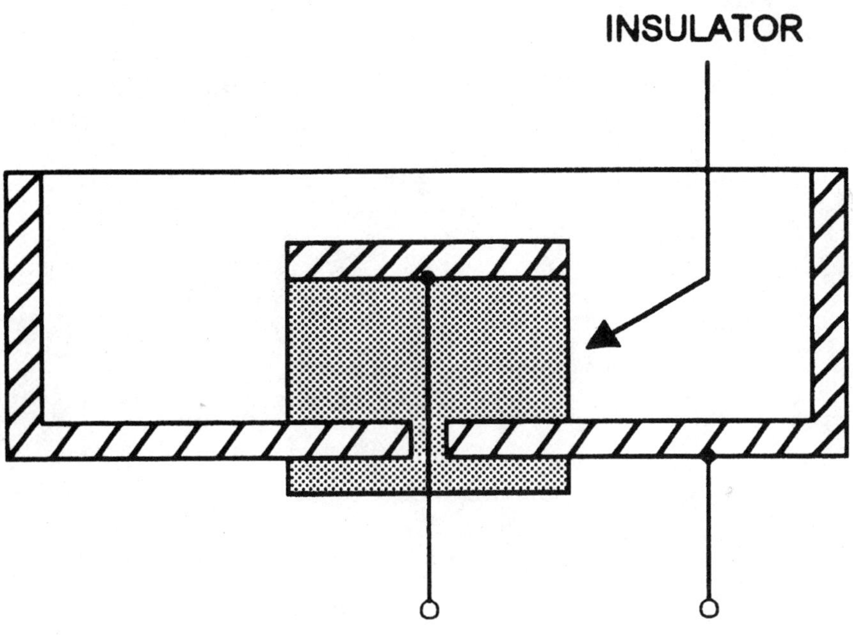

Figure 10-1d. Tank level detector.

Chapter 11

Missing Pulse Detector

A missing pulse detector circuit remains dormant as long as a series of trigger pulses are received, but will produce an output pulse when an expected pulse is missing. These circuits are used in a variety of applications including alarms. For example, in a bottling plant softdrink cans are packaged into six-packs. As each can passes a photocell a pulse is generated to the input of a missing pulse detector. If a pulse is not received, however, the machine knows that the count is one can short, so issues an alarm or corrective action. Similar circuits can be used in a wildlife photography system. An infrared light emitting diode (LED) is modulated or chopped with a pulse waveform. As long as the pulse is received at the sensor, the circuit is dormant. But if an animal passes through the IR beam even briefly a missing pulse detector will sense its presence and issue and output that fires a camera flashgun and electrical shutter control.

Figure 11-1a shows the circuit for a missing pulse detector based on the 555 IC timer, while **Figure 11-1b** shows the timing waveforms. This circuit is the standard 555 MMV, except that a discharge transistor is shunted across capacitor C1. When a pulse is applied to the input it will

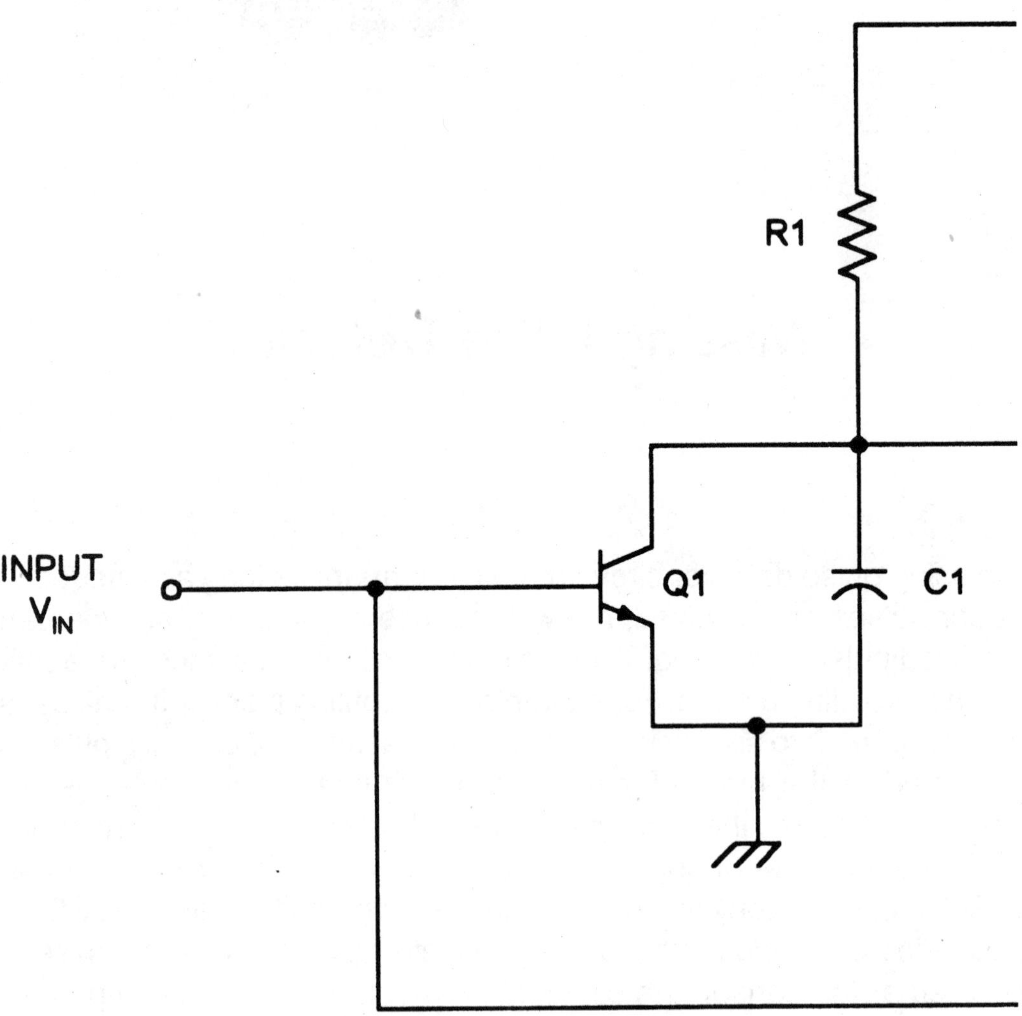

Figure 11-1a. Missing pulse detector.

trigger the 555 and turn on Q1, causing the capacitor to discharge. After the first input pulse the output of the 555 snaps HIGH and remains HIGH until a missing pulse is detected.

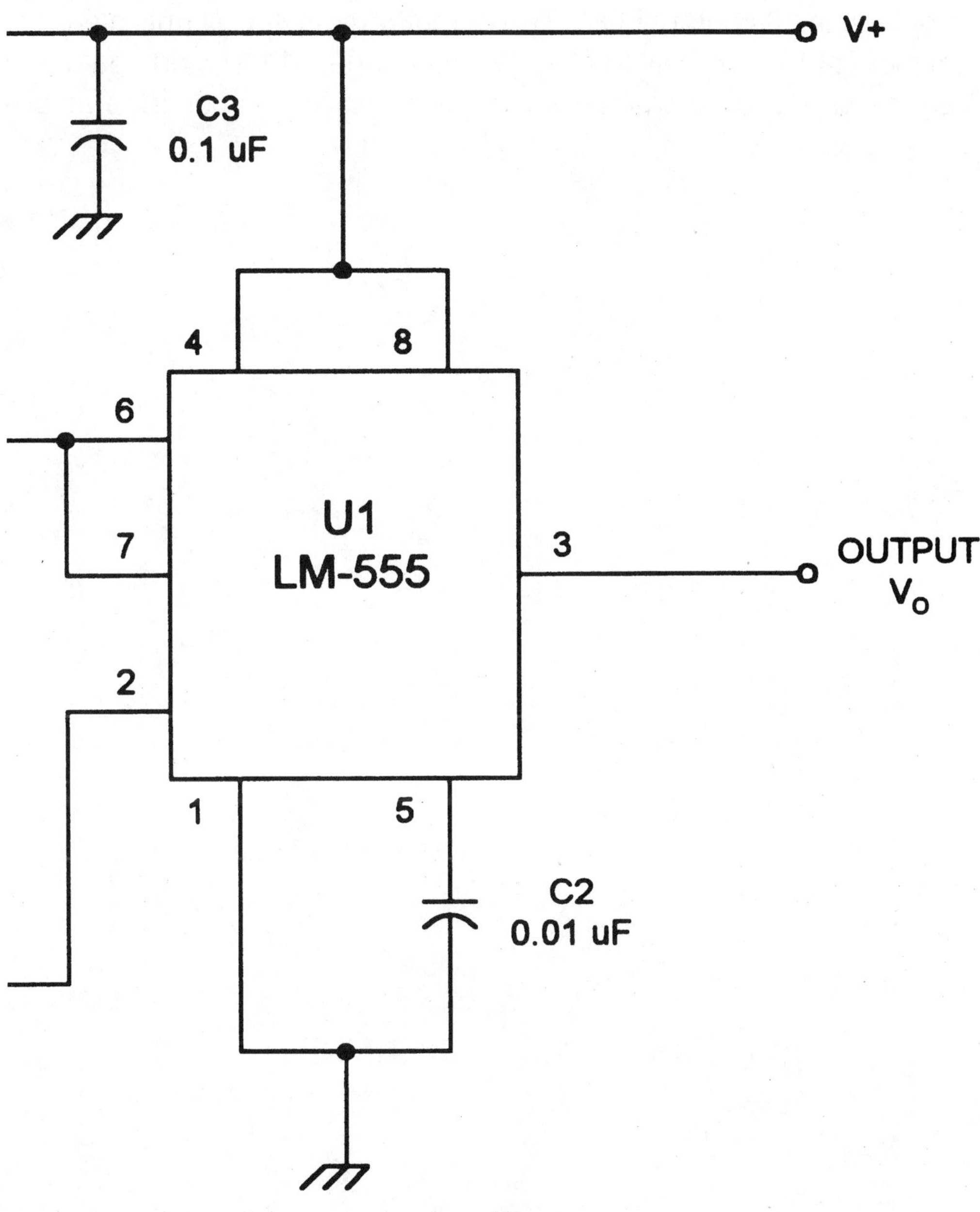

Circuit action can be seen in **Figure 11-1b**. At times T1 and T2 input pulses are received. As long as (T2-T1) is less than the time required for C1 to charge up to 2(V+)/3 the 555 will never time-out. But if a pulse is missing, as at T3, the capacitor voltage continues to rise to the critical 2(V+)/3 threshold value. When V_c reaches this point the 555 will time-out forcing its output LOW. The output remains LOW until a subsequent input pulse is received (T4), at which time Q1 turns on again and forces the capacitor to discharge. The cycle can then continue as before.

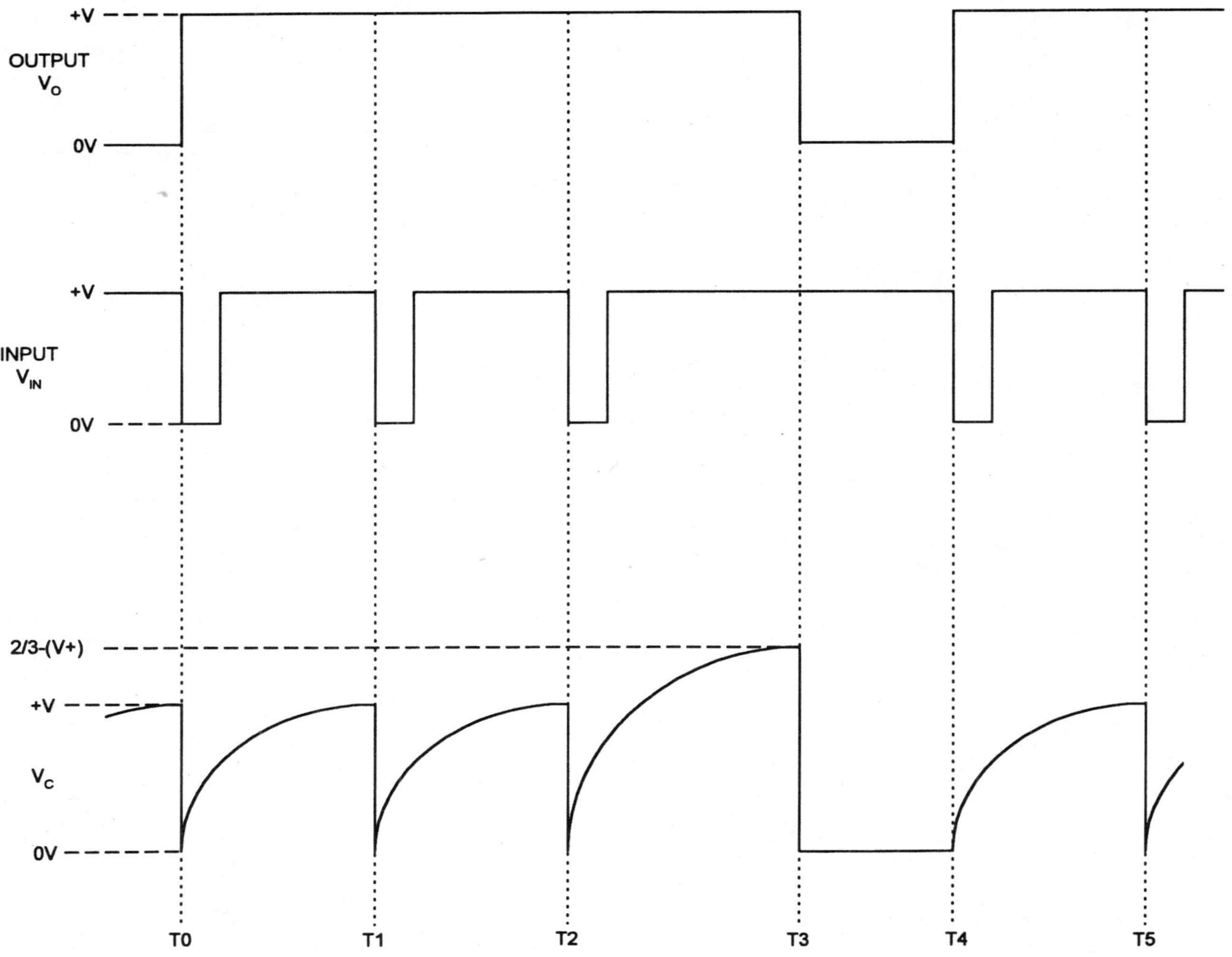

Figure 11-1b. Timing waveforms.

Chapter 12

Pulse Position Circuit

A pulse positioner is a circuit that will allow adjustment of the timing of a pulse to coincide with some external event. For example, in some instrumentation circuits a short pulse must be positioned to a certain point on a sine wave (e.g., the peak). The pulse positioner could be triggered from the zero-crossing of the sine wave, and then adjusted to place the output pulse where it is needed.

The circuit is shown in **Figure 12-1a**. **Figure 12-1b** shows the concept of pulse positioning using two one-shot circuits, labelled OS1 and OS2. The re-positioned pulse is not actually the original pulse, but rather it is a recreated pulse with similar characteristics. The input pulse is used to trigger OS1. The duration of this one-shot circuit is fixed to the delay required of the re-positioned pulse. If the delay must be variable, then resistor R1 is made variable. When OS1 times-out it will trigger OS2.

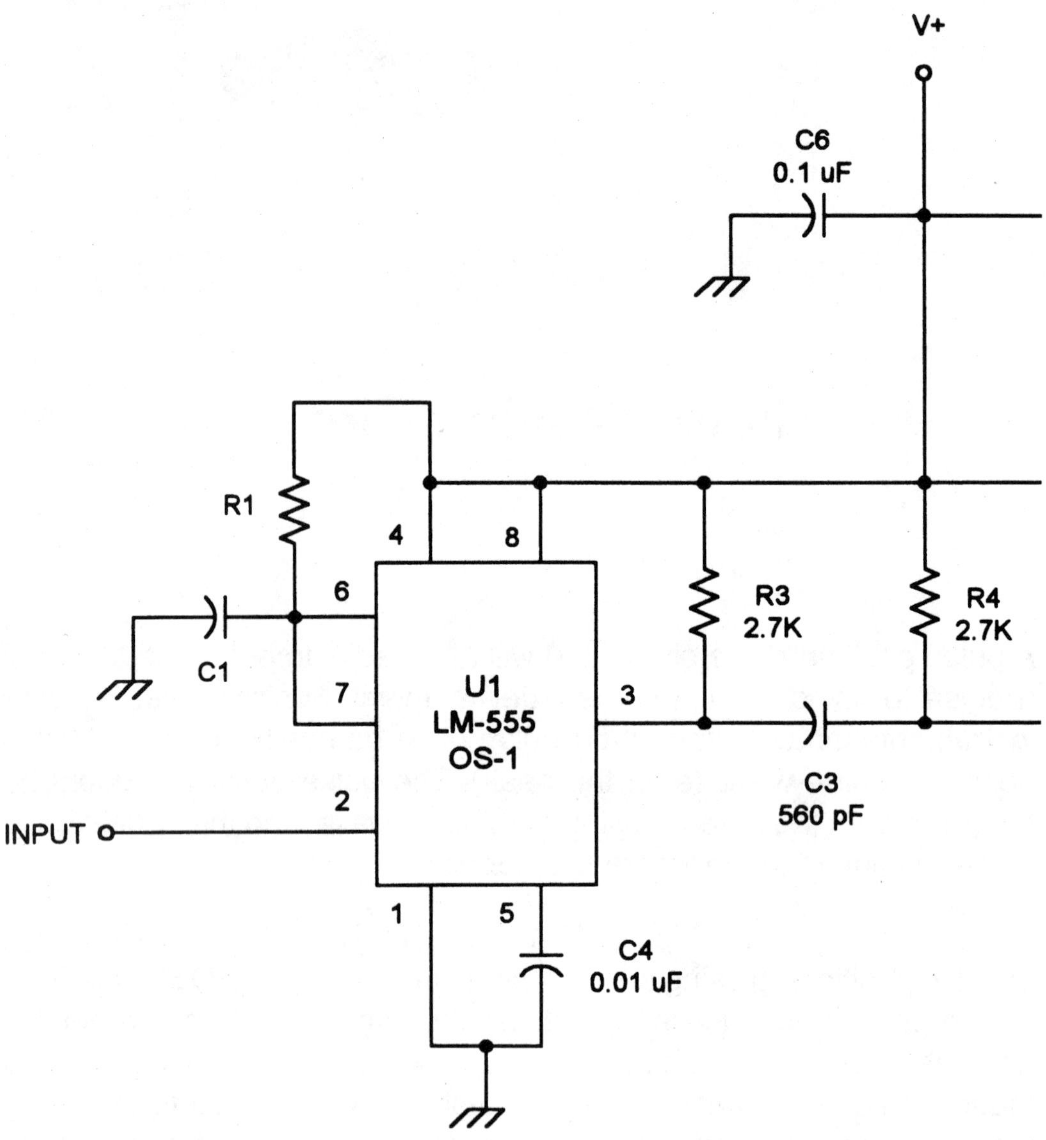

Figure 12-1a. Pulse positioning circuit.

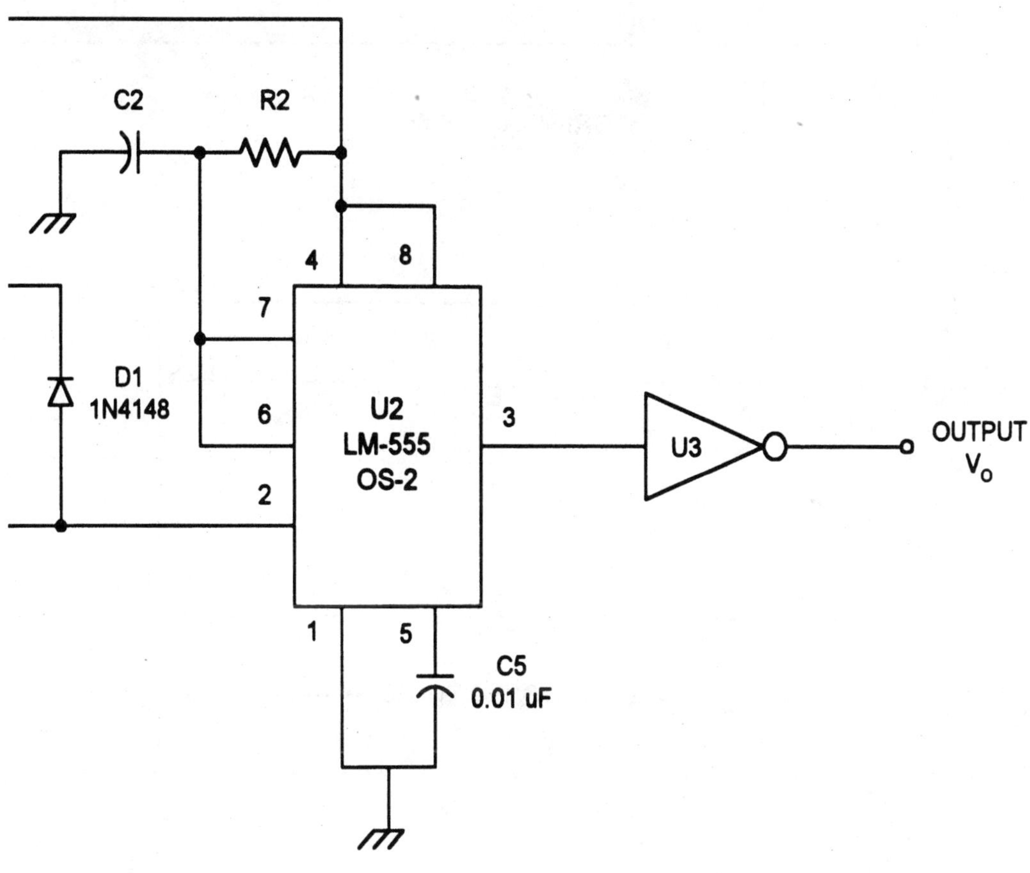
C2
R2
4
8
7
D1
1N4148
6
U2
LM-555
OS-2
3
U3
OUTPUT
V_O
2
1
5
C5
0.01 uF

The output pulse of OS2 is set to the parameters of the original input pulse. An inverter circuit is used to make the output of OS2 have the same polarity as the trigger pulse at the input of OS1. To an outside observer the pulse appears to have been re-positioned, although in fact it was merely recreated at time *T* (the delay period in **Figure 12-1b**).

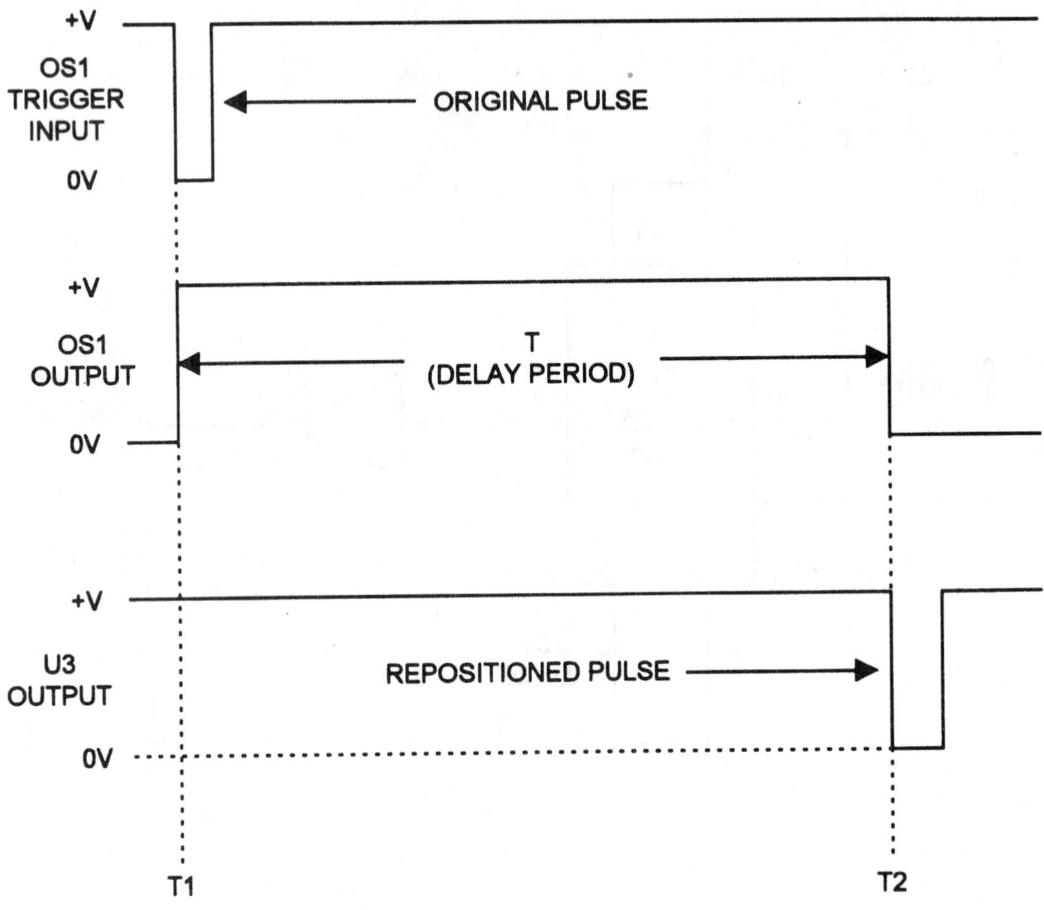

Figure 12-1b. Pulse positioning waveform.

Chapter 13

Tachometry Using a One-Shot Circuit

The word tachometry is used to designate the measurement of a repetition rate. In the automotive tachometer, for example, the instrument counts the pulses produced by the ignition coil to measure the engine speed in revolutions per minute (RPM). In medical instruments it is often necessary to measure factors such as heart or respiration rate electronically using tachometry circuits. A heart rate meter (cardiotachometer) measures the heart rate in beats per minute (BPM), while the respiration meter (pneumotachometer) measures breathing rate in breaths per second.

There is a certain commonality among non-digital tachometer circuits. It doesn't matter whether the rate is audio or sub-audio, or even above the audio rate, the basic circuit design is the same.

Figure 13-1 shows the basic tachometer circuit in block diagram form. Not all of the stages will be present in all circuits, but some of them are basic to the problem so are universally found. The AC amplifier and Schmitt trigger will be used only when needed. These stages are used for input signal conditioning, so are used only where such conditioning is needed. The one-shot circuit and the Miller integrator are basic to the design, however, so are used for all such circuits.

The idea is to convert a frequency or repetition rate to an analog voltage. This is done by first converting the signal to pulse form. The AC input amplifier is used only if it is necessary to scale the input signal to a level where it will drive a Schmitt trigger or other squaring circuit. The purpose of the following stage is to produce a square wave output signal at the same frequency of as the input signal.

The purpose of the stages in **Figure 13-1** is to produce a DC voltage output that is proportional to the input frequency or pulse repetition rate. The integrator is designed to produce an output voltage that is the time-average of the input signal. That is, the integrator output is proportional to the area under the input signal. The job of the tachometer designer is to create a situation in which the only variable is the frequency or repetition rate of the input signal. Variation in other factors obscures the results.

The output pulse of a one-shot circuit has a constant amplitude and constant duration. The area under the pulse is the product of the amplitude and duration, so from pulse to pulse the area does not

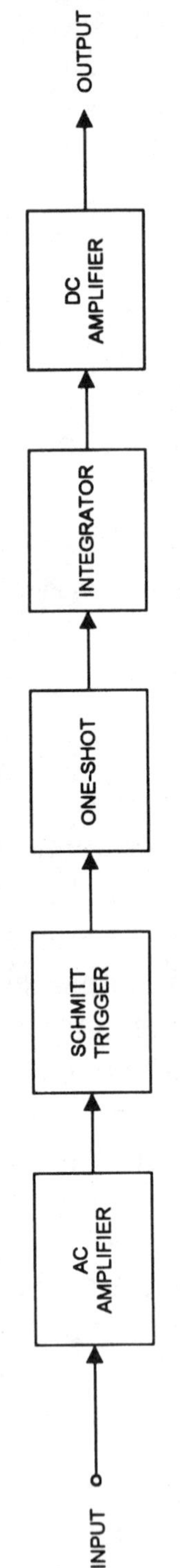

Figure 13-1. Tachometer block diagram.

change. If the one-shot is constantly retriggered by the input signal, the total area under the resultant pulse train is a function of only the number of pulses. Therefore, the time-average of the integrator output will be a DC voltage that is proportional to the input frequency.

Figure 13-2a shows a practical application of the tachometer principle. This circuit was used to demodulate the audio frequency modulated signal from an instrumentation telemetry set. A similar circuit (but not based on the 555) was once popular as a coilless FM detector in communications and broadcast receivers. These pulse counting detectors operated at 10.7 MHz (a commonly used FM IF frequency in receivers). The circuit shown in **Figure 13-2a** was used to demodulate a human electrocardiograph (ECG) signal transmitted over telephone lines. The ECG is an analog voltage waveform, and was used to frequency modulate an audio voltage controlled oscillator (VCO) at the transmit end. Normally, the ECG has too low a Fourier frequency content (0.05 Hz to 100 Hz) to pass over the restricted passband of the telephone lines (300 Hz to 3000 Hz). But when used to frequency modulate a 1500 Hz carrier, however, the signal passed easily over telephone circuits.

The circuit for the demodulator circuit is shown in **Figure 13-2a**. The input waveshaping function is performed by an LM-311 voltage comparator. The job of the LM-311 is to square the 200 mV peak-to-peak sine wave input signal so that it is capable of triggering the 555 (U2). In this mode the LM-311 is operating basically as a zero-crossing detector circuit.

The output of the 555 is a pulse train that has constant amplitude and duration. These pulses vary only in repetition rate, which is the same as the frequency of the input signal. The 555 output pulses are integrated in a passive RC integrator (R5-R7/C4-C6). The output of the integrator is a DC voltage that is a linear function of input frequency (see **Figure 13-2b**). This DC voltage can be scaled, if necessary, to any desired level.

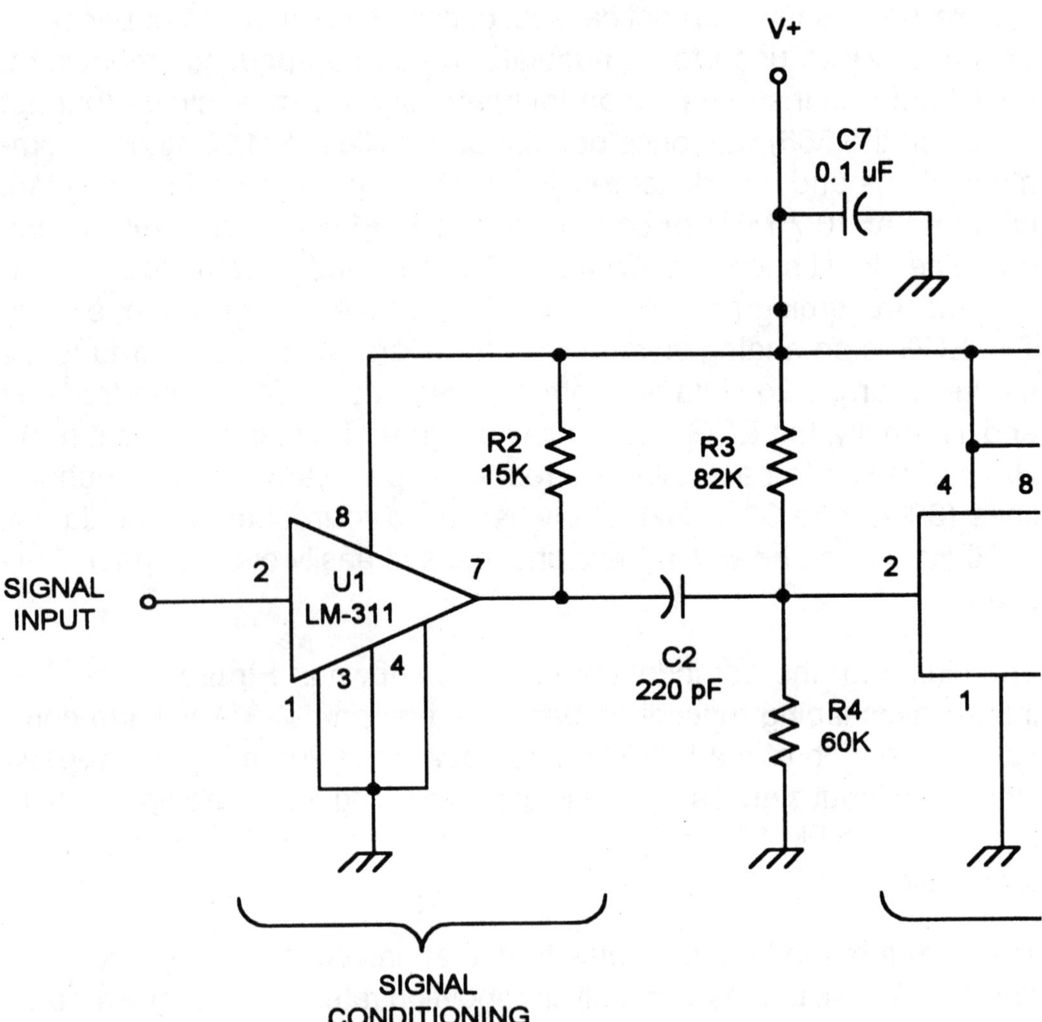

Figure 13-2a. Tachometer circuit.

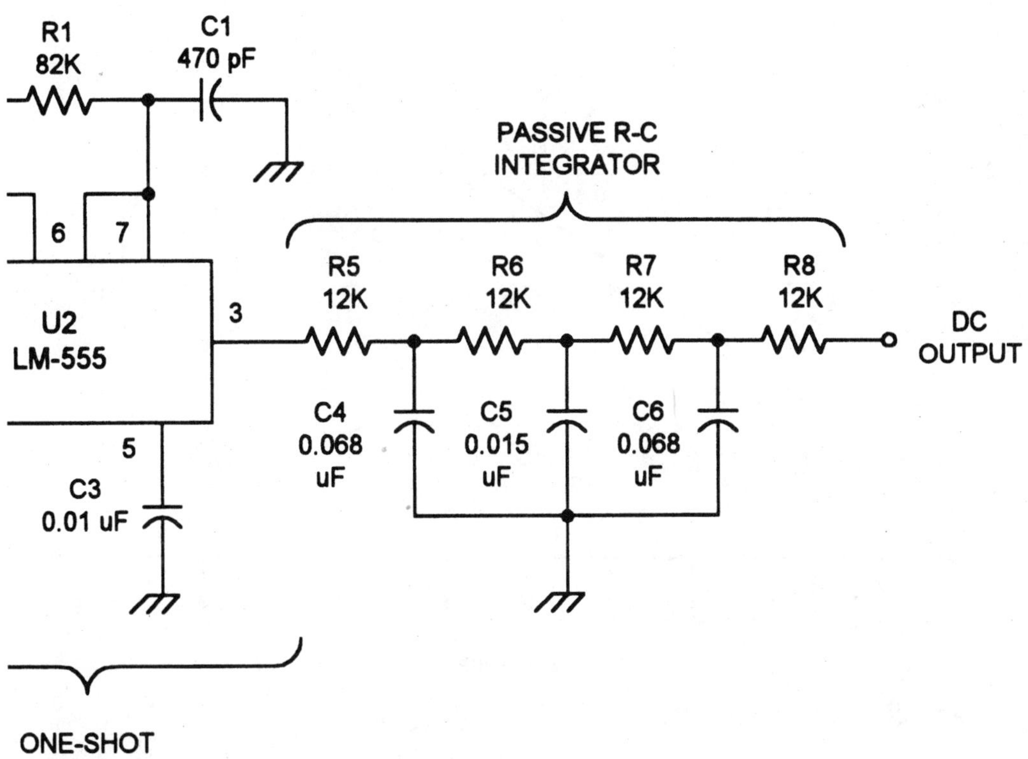
R1
82K
C1
470 pF
PASSIVE R-C
INTEGRATOR
6
7
U2
LM-555
3
R5
12K
R6
12K
R7
12K
R8
12K
DC
OUTPUT
5
C3
0.01 uF
C4
0.068
uF
C5
0.015
uF
C6
0.068
uF
ONE-SHOT

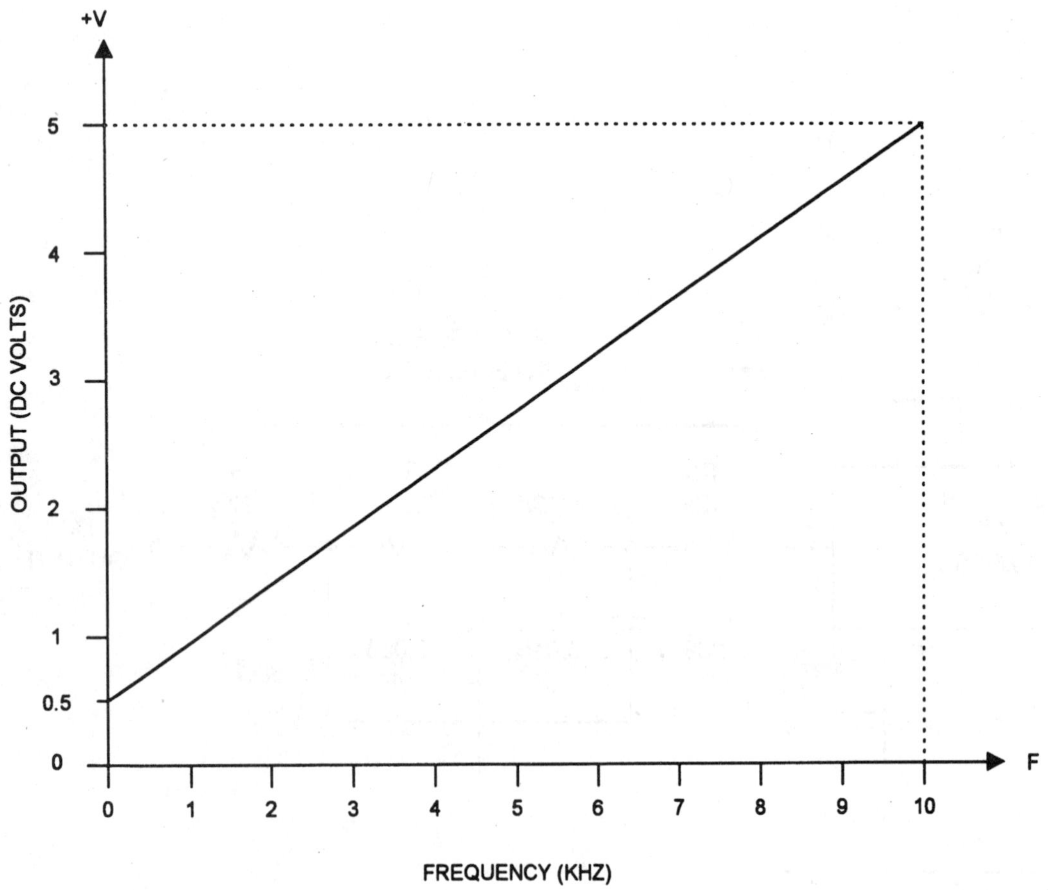

Figure 13-2b. Output function.

Chapter 14

Analog Audio Frequency Meter

A related circuit is shown in **Figure 14-1**. This 555-based tachometer is used to measure audio frequency over four ranges: DC to 50 Hz, DC to 500 Hz, DC to 5,000 Hz and DC to 50,000 Hz. The circuit uses the same form of input signal conditioning as the previous circuit, and uses a 555 as the one-shot circuit. The integration function is taken up by the combination of RC network R4/C4 and the mechanical inertia of the meter (M1) movement.

The variable potentiometer can be used to calibrate the meter. Also, you can use either R1 or C1 as variable elements for calibration of the meter movement. The output is linear, and so is the meter. Given that the meter movement has a full-scale reading of 50 µA, the "5" at the upper end of each range and the meter movement can be used as calibration points without the need for reprinting the meter movement, or using a conversion chart.

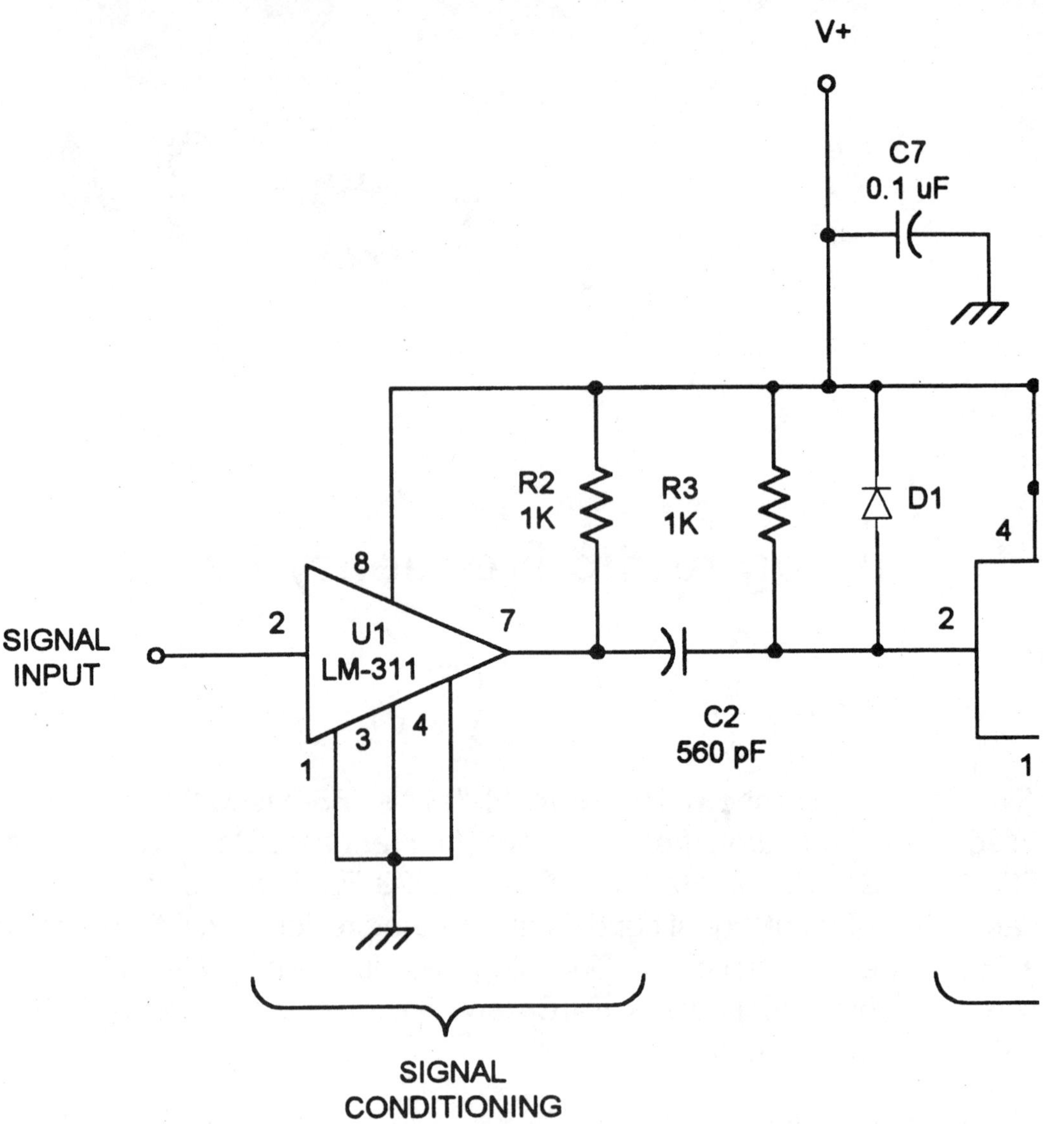

Figure 14-1. Analog frequency meter.

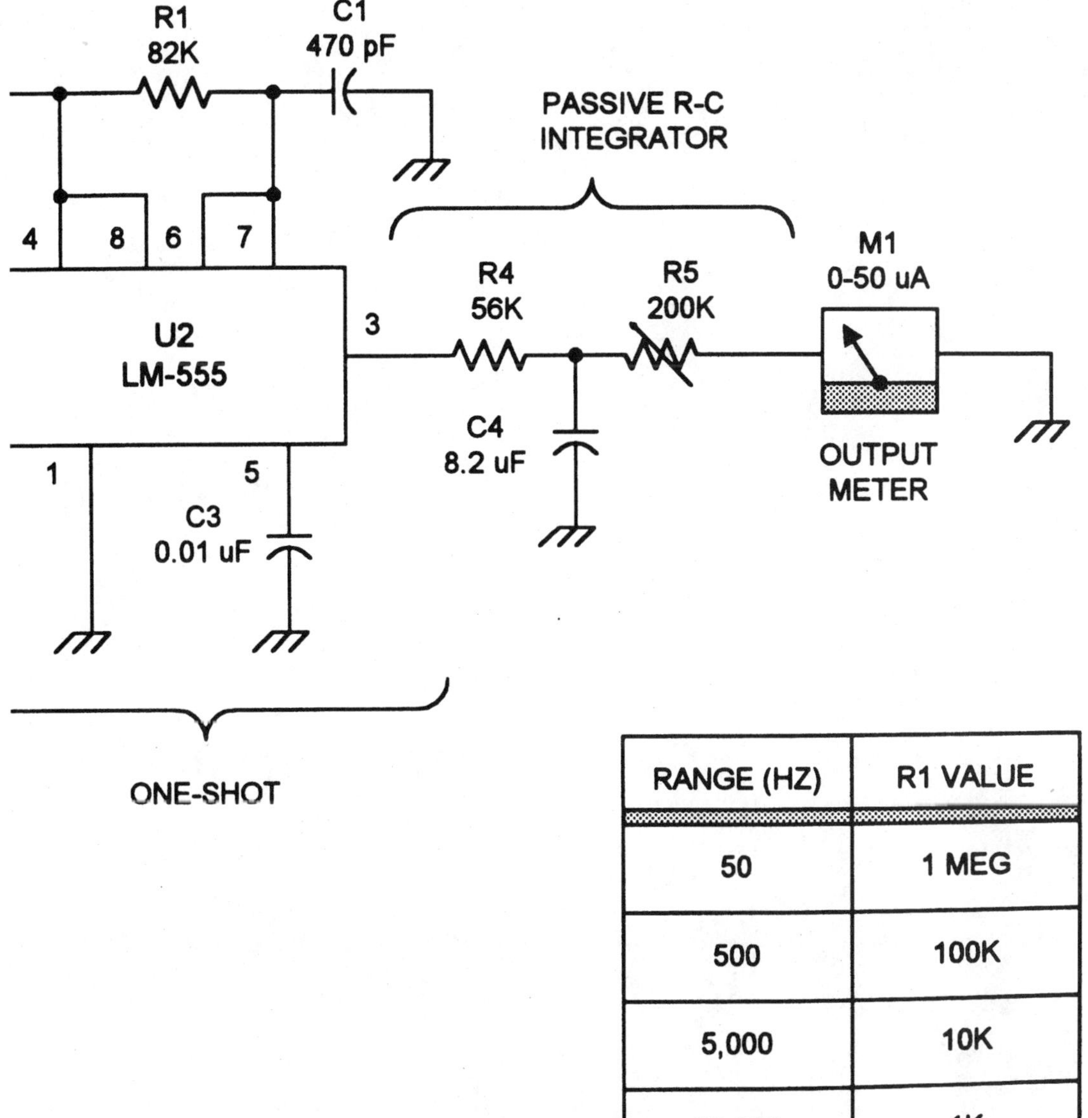

RANGE (HZ)	R1 VALUE
50	1 MEG
500	100K
5,000	10K
50,000	1K

Chapter 15

One-Second Timer/Flasher

There are many uses for a one-second timer/flasher. If you build a digital frequency counter, for example, you would use the one-second timer to control the display update line. The display is kept stable by only displaying immediate past "old" data while the new data is being accumulated. The LED/LCD display numerals do not show the count being totalized.

In the one-second timer of **Figure 15-1** we connect the output to a light emitting diode (LED) that will blink on and off at a one-second rate. The LED is good for demonstration, and could be used for the update light on a counter. You can also connect the output (pin no. 3) to any other circuit you wish.

The circuit is an ordinary LM-555 monostable multivibrator. The time-constant (R1 + R2) x C1 is adjusted to give the one-second period required. Repetitive blinking requires repetitive triggering.

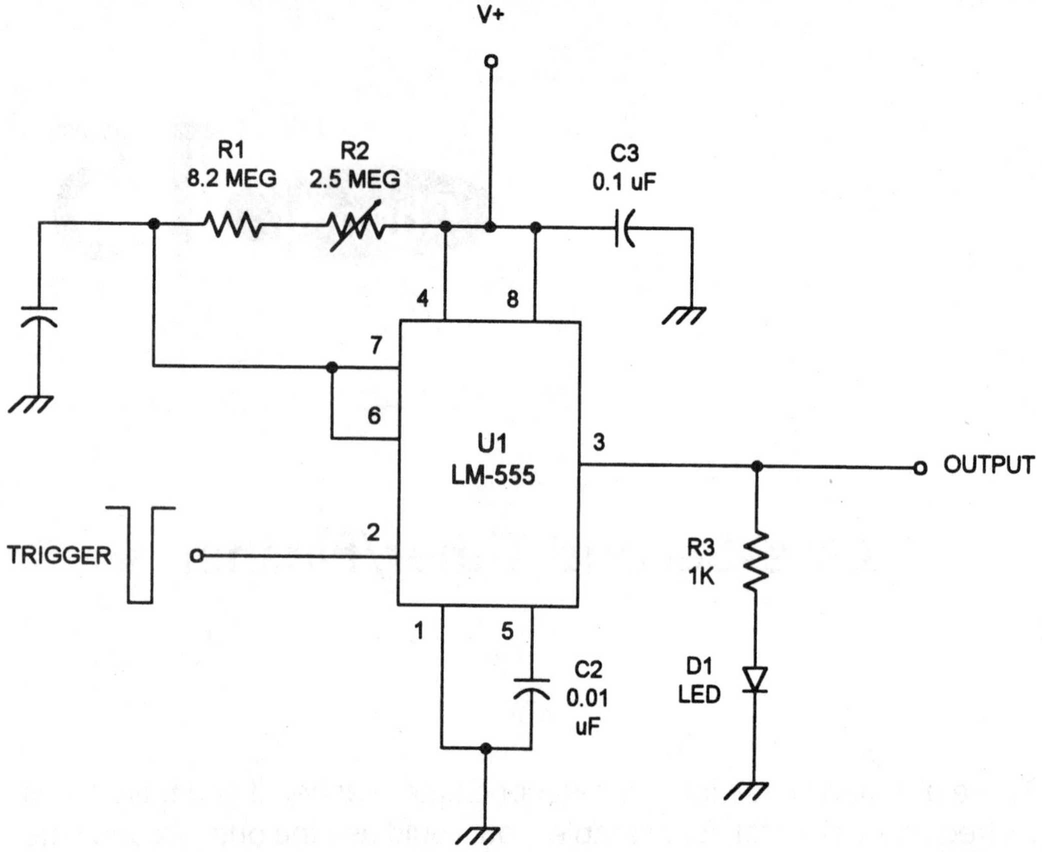

Figure 15-1. Variation pulse width one-shot.

Note that the resistance of the circuit is split between two resistors, while the capacitance of the timing circuit is a single capacitor (switch selecting different capacitors allows different timing ranges). It is relatively easier to find precise standard values of capacitance, provided that you can live with a standard value for the given application. Once the capacitance is selected you can calculate the required resistance.

The timing resistor is made by connecting a fixed resistance in series with an adjustable resistance. This procedure sets a minimum and maximum timing duration.

Chapter 16

Level Translation of LM-555 Output

The LM-555 output levels are zero volts and a positive voltage set by the V+ voltage applied to pin no. 8. Sometimes it is necessary to drive a circuit with a different level. For example, suppose an LM-555 timer is operated from a +12 VDC power supply, but must interface with TTL logic devices.

The LM-555 output will either sink or source currents up to 200 mA, so it will be directly compatible with TTL devices if the LM-555 is operated from a +5 VDC power supply. No additional circuitry is needed in that case. But what about the case where the LM-555 device is operated from DC power supply voltages other than +5 VDC? Remember, the LM-555 will allow operation over the range +4.5 to +18 VDC.

Method I

Figure 16-1a shows the use of a bipolar NPN transistor to make the level translation. The collector voltage and circuit configuration suggest that this circuit will drive a TTL input. The output from the transistor is inverted relative to the LM-555 output polarity; i.e., if the LM-555 output is LOW, the output of the transistor is HIGH (and vice versa). Almost any NPN transistor will work for Q1 (e.g., 2N2907, 2N3393, 2N3904). The collector of the transistor is connected to a +5 VDC source (in this case the output of a +5 volt three-terminal regulator) through load resistor R2. The size of this resistor is selected to limit the collector current to a value that is within the transistor's limitations, although the value shown is nearly universal for TTL circuits with fan-out up to ten.

When the output of the LM-555 is LOW, the transistor is unbiased and no collector current will flow. This makes the collector terminal of the transistor +5 volts, which is TTL HIGH. But when the output of the LM-555 is HIGH, the transistor becomes forward-biased. The base resistor is selected to permit a base current that drives the transistor into saturation. This situation drives the collector LOW, i.e., to below the saturation voltage ($V_{CE(SAT)}$) of the transistor.

Method II

Practical experience with TTL devices reveals a basic problem: the output terminal (pin no. 3) is not well isolated from the triggering circuitry inside. A pulse applied to the output terminal will cause triggering of the LM-555. This can occur when the device is used to drive a relay coil. We can isolate the output of circuits such as **Figure 16-1a** by connecting several diodes in series with the base of the transistor (see **Figure 16-1b**). The total junction potentials of the diodes must be sufficient to snub any pulse applied to the output.

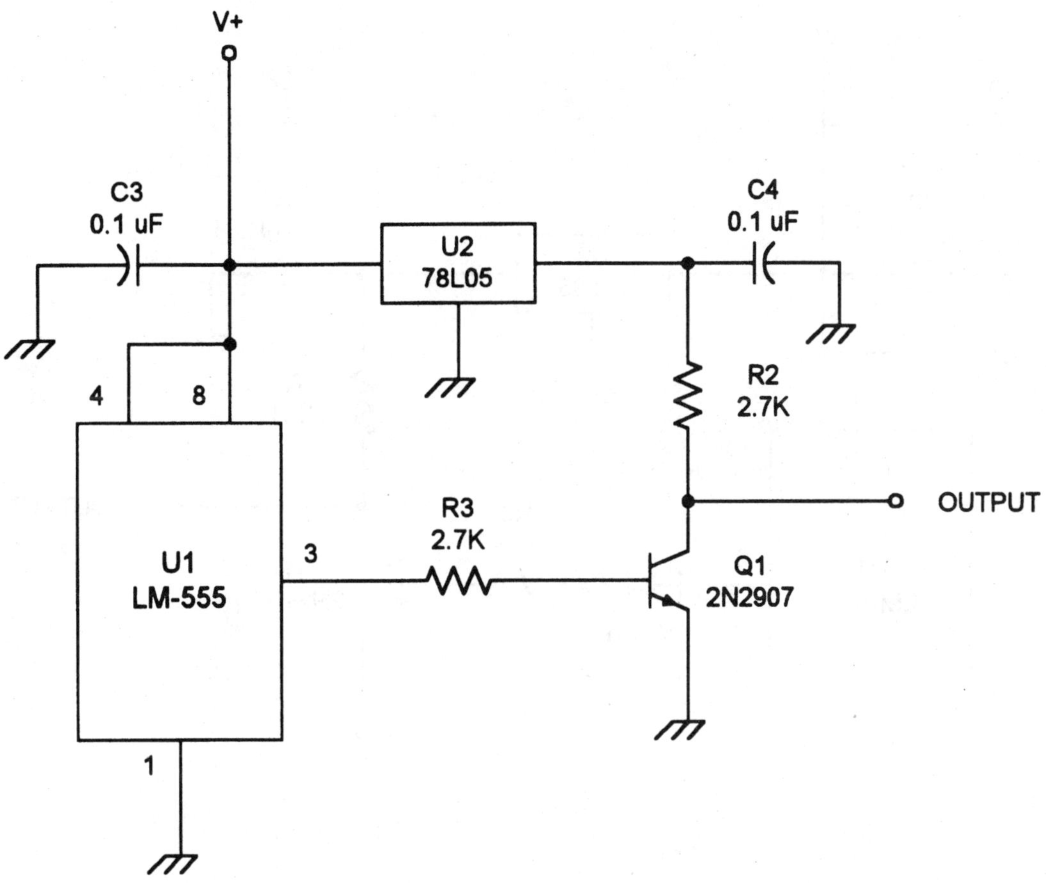

Figure 16-1a. Level translated output.

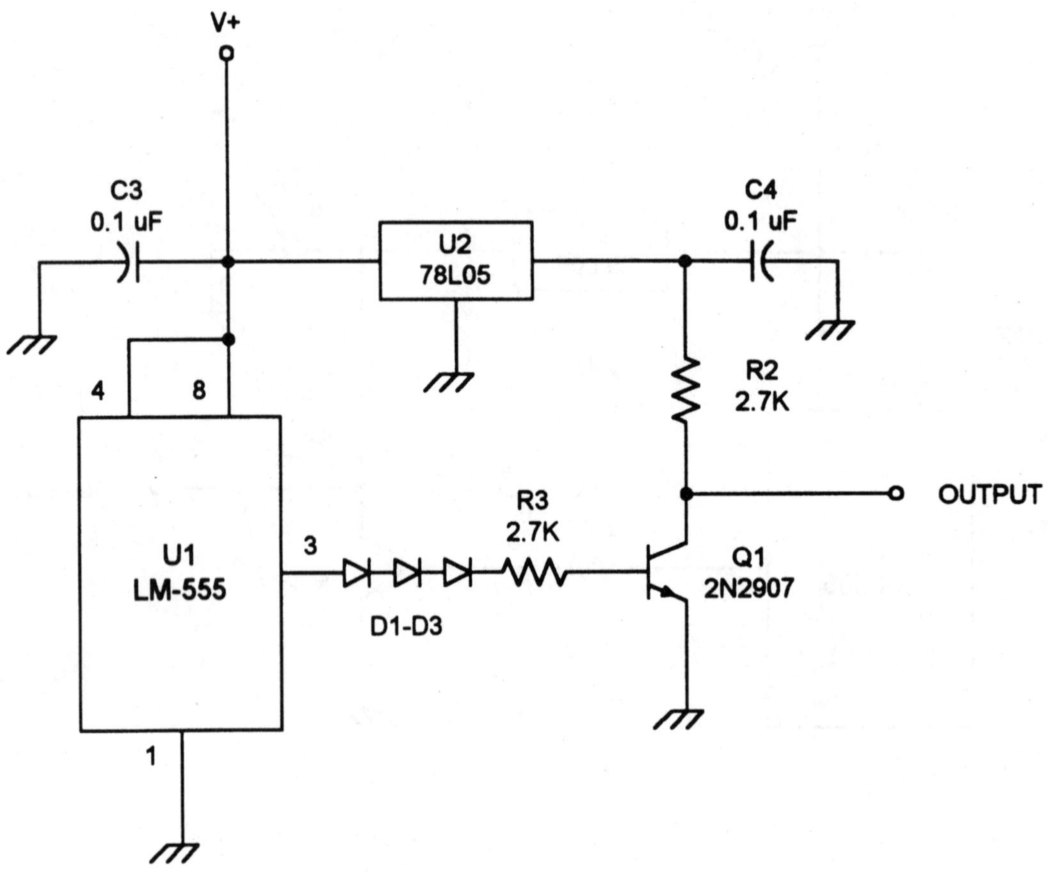

Figure 16-1b. Diode protected level translated output.

Method III

The LM-555 supports pretty much the same range of operating potentials as the CMOS line of logic. Because CMOS input impedances are very high, the LM-555 is not taxed at all driving CMOS devices.

There are two related CMOS devices that are of immediate interest here: 4049 and 4050. Both of these will accept CMOS input levels and (if a +5 VDC supply is used) produce TTL outputs. The 4049 device is a set of six inverters in a single package. The output stage of each inverter is a TTL-compatible circuit. The 4050 is similar, except that it is a hex buffer. A "buffer" provides no inverting action, i.e., HIGH in produces HIGH out, and LOW in produces LOW out. We can use either of these devices to invert the LM-555 timer circuits to TTL logic circuits. An example is shown in **Figure 16-1c**.

One section of a 4049 hex inverter is connected between the output of the LM-555 and the load. When the output of the LM-555 is LOW, then the output of the 4049 is HIGH; when the output of the LM-555 is HIGH the inverter output is LOW.

Exactly the opposite situation is seen when a 4050 is used instead of an inverter: a LOW LM-555 output produces a LOW 4050 output, and a HIGH LM-555 output produces a HIGH 4050 output.

Because CMOS devices so lightly load the LM-555 output, it is possible to put several 4049 or 4050 devices in the circuit to produce multiple outputs. For example, using a 4049 section and a 4050 section, with their inputs in parallel and connected to the LM-555 output, produces complementary HIGH/LOW outputs.

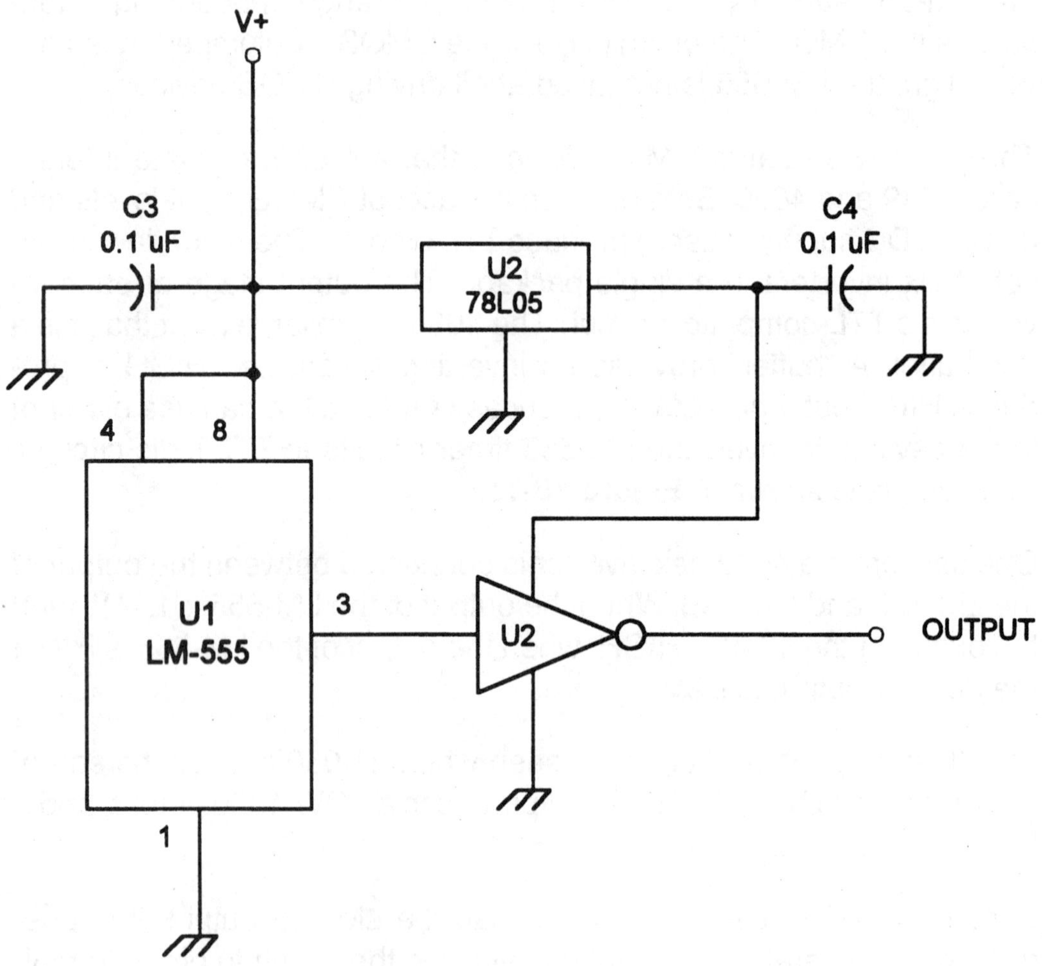

Figure 16-1c. Using an inverter output.

Chapter 17

100-kHz Oscillator

Figure 17-1 shows the circuit for a 100 kHz oscillator based on the LM-555 astable multivibrator configuration. Because the LM-555 will sink or source up to 200 mA of current, this circuit is capable of directly driving a small loudspeaker or earphone.

This circuit can also be used as a troubleshooting "signal squirter" for AM radios and audio equipment. In that application delete the loudspeaker and R3, and connect the output of the LM-555 through a 0.01 μF capacitor to the circuit under test. Because the output of the circuit is a square wave it will be rich in harmonic content. The 1,000 Hz square waves produce usable harmonics well into the medium wave band, so they can be used to troubleshoot not only AM band radios, but also other RF circuits up to about 5 MHz (5,000 kHz). The signals appear as "birdies" every 1 kHz along the spectrum.

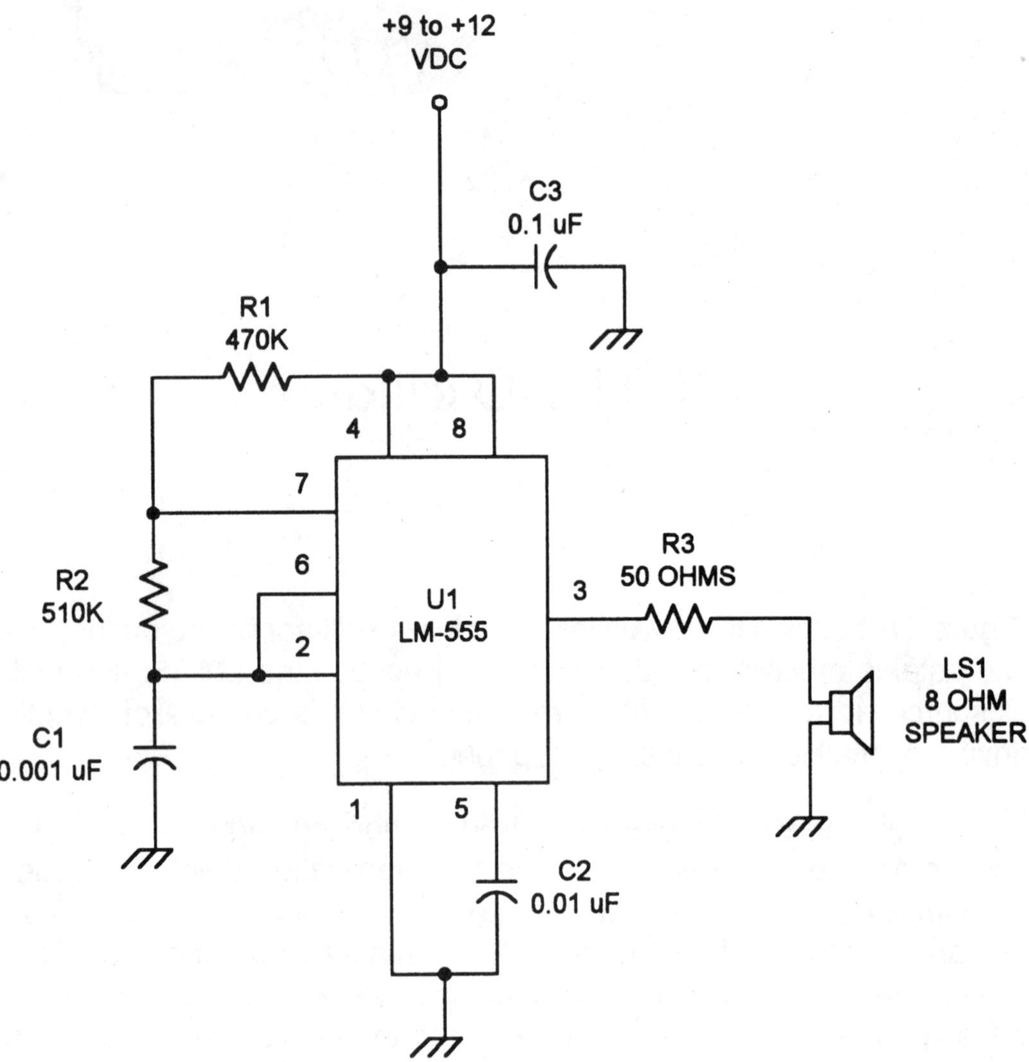

Figure 17-1. Audio tone generator.

Chapter 18

Relay and Optoisolator Drives

There are times when you simply don't want to connect an LM-555 or other IC timer to a particular load circuit. For example, when 120 volts AC is being switched, or where the load involves a high voltage DC potential. These and other situations call for isolation between the IC and the load. Sometimes, the reason is safety of personnel, while in other cases the reason is to prevent damage to the IC device being used.

A relay is an electromechanical switch that uses an electromagnetic coil to cause switching action. The relay shown in **Figure 18-1a** has a set of single-pole, double-throw contacts (A1,A2 and A3), or "SPDT" as it is often called. Other forms (SPST, DPDT SPDT) are also available.

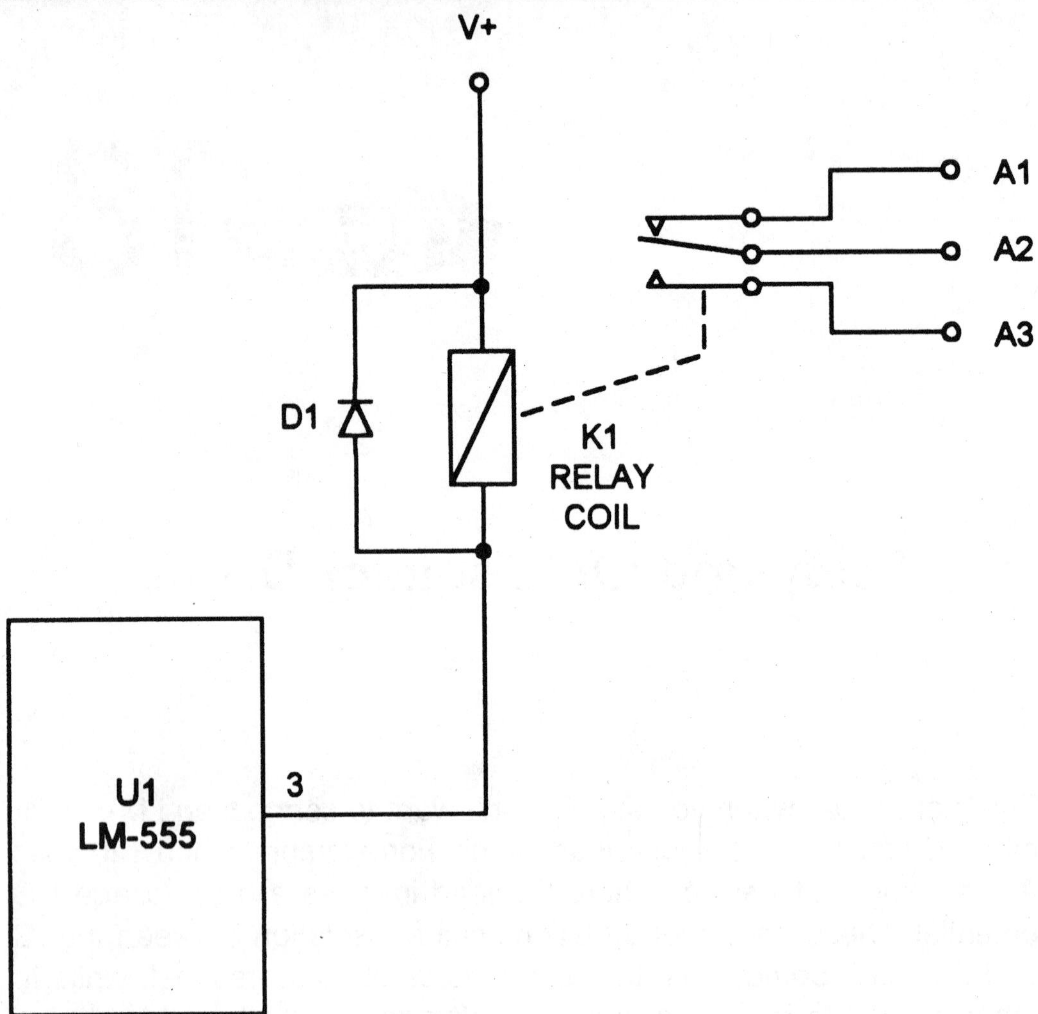

Figure 18-1a. Simple relay interface.

The contacts of relay K1 (A1-A3) are shown in the deenergized (i.e., no current in the K1 coil) position, with A1-A2 shorted and A3 open. In normal schematic diagram practice the deenergized position of the contacts is assumed unless noted otherwise. When a current flows in the coil, producing a magnetic field around the coil, the contacts pull to the energized position (which shorts A2-A3, leaving A1 open).

The LM-555 output (pin no. 3) will sink or source up to 200 mA of current. In **Figure 18-1a** the relay is connected between the V+ DC voltage and the output terminal, so the LM-555 is being asked to sink the coil current. As long as the coil current is less than 0.2 amperes (200 mA), then this circuit will work. Unfortunately, relay coils often draw considerable current, so this method works for small relays and reed relays only. At +12 VDC, this circuit will work with coils that have DC resistances of 60 ohms or more.

Note that there is a diode across the K1 relay coil. This diode is called a "snubber diode" and is absolutely critical in solid-state circuits where relays are used. It is used to snub the counterelectromotive force (CEMF) produced by the relay when it deenergizes. When current is applied to the coil, a magnetic field is built up, which stores energy. When the current ceases to flow, the magnetic field collapses suddenly causing a current to be induced into the turns of the relay coil with the opposite polarity of the energizing current. Because the induced CEMF, a voltage, is proportional to both the coil inductance and the time rate of change of the current [$V = L(\Delta I/\Delta t)$], which at turn/off is very rapid, the CEMF is very high. Thus, the deenergizing relay produces a very high voltage that is of the opposite polarity from V+.

When the circuit is dormant, and during turn-on of K1, diode D1 is reverse-biased, so it is not in operation. When the relay deenergizes, creating the CEMF spike, then the diode is forward-biased and will drop the CEMF to the junction potential of the diode (typically 0.6 to 0.7 volts for silicon).

Early in the solid-state era a lot of aircraft and automobile electronics equipment suffered odd failures because the relays and solenoids used for electromechanical actuators produced killer transients when turned off. The problem became especially acute when the digital age arrived for vehicles in the early 1980s. Digital circuits may or may not be destroyed by the transients, but they often accept that transient as a valid digital signal...and trip off into never-never land. After that problem was diagnosed, relays and solenoids aboard automobiles and airplanes began to incorporate snubber diodes (some relays have them built-in).

Because the LM-555 will either sink or source current, the relay can also be connected directly between the output and ground. In that case, V+ is not involved and the LM-555 supplies (sources) the coil current. Where **Figure 18-1a** actuates the relay when the LM-555 output is LOW, the sourced connection version actuates when the output is HIGH.

Another approach is shown in **Figure 18-1b**. In this circuit one section of a TTL 7406 hex inverter is used as the relay driver. The 7406 device is an open-collector circuit, so it can be connected to the relay as shown. The advantage of the 7406 is that is operates from TTL DC voltages,

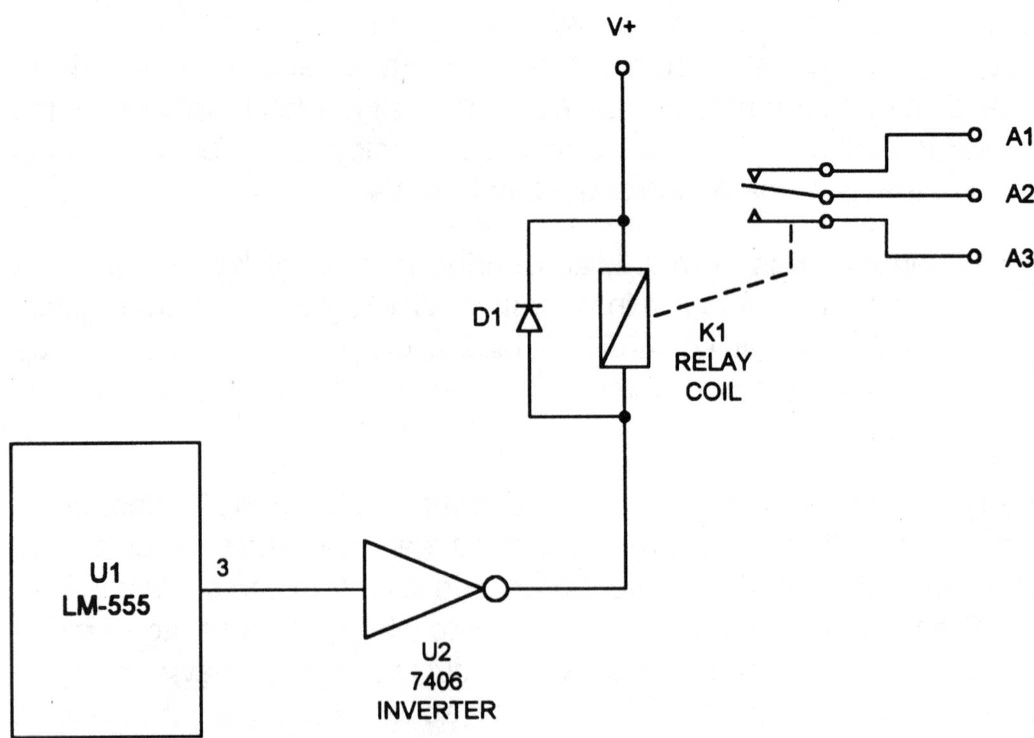

Figure 18-1b. Inverter interfaced relay.

but the output transistor can work at potentials up to +30 volts DC. The inverter causes the relay to energize when the output of the LM-555 is HIGH, making the 7406 output LOW.

The circuit in **Figure 18-1c** is similar to **Figure 18-1b**, but has a self-latching feature. A second set of SPDT relay contacts (B1-B3) are required on the relay. In the deenergized position (shown) B1-B2 are shorted together and B3 is open. When the relay is energized, B2-B3 are shorted and B1 is open. This situation puts the cold end of coil K1 at ground potential, holding it energized even after the output of the LM-555 goes LOW again. To deenergize the relay press the normally-open reset switch (S1).

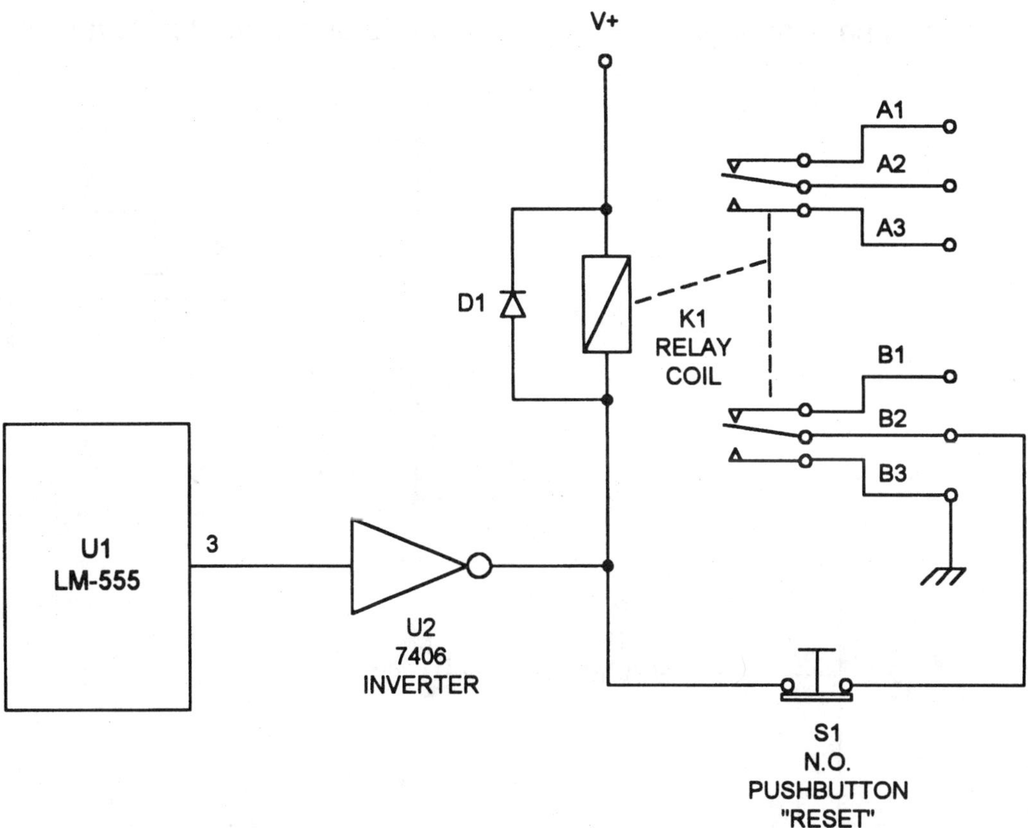

Figure 18-1c. Self-latching relay.

A transistor relay driver is shown in **Figure 18-1d**. In this circuit a transistor (Q1) is connected so that its base is driven into conduction when the output of the LM-555 output is HIGH. This creates a low resistance path between the collector and emitter of Q1 because the transistor is saturated, and therefore energizes the relay coil. This circuit will work for any transistor that can be saturated at base currents of 200 mA or less, which means it can drive some hefty collector currents (which is the relay coil current), depending on the beta gain of the transistor.

In cases where the gain is insufficient, then you might want to use a Darlington amplifier driver as shown in **Figure 18-1e**. Although Darlington devices can be bought, they can also be made using a pair of transistors. The overall gain of a Darlington pair is the product of the individual gains of Q1 and Q2. For example, if both Q1 and Q2 are identical devices, and have beta gains of 100, then the total apparent beta gain of

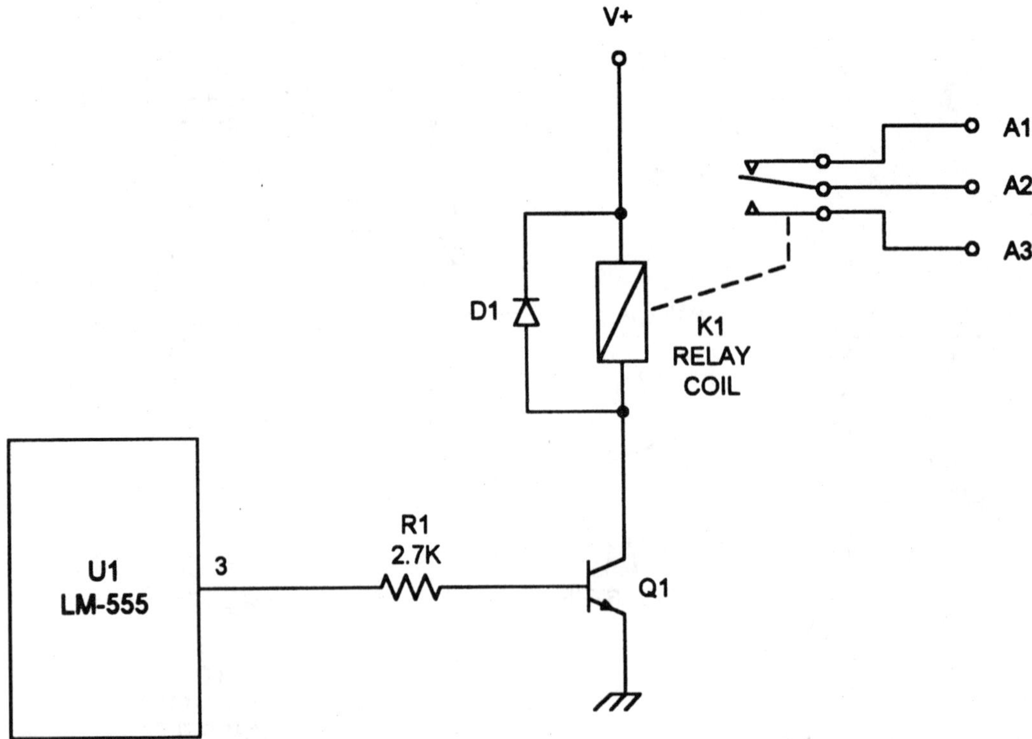

Figure 18-1d. High-current relay interface.

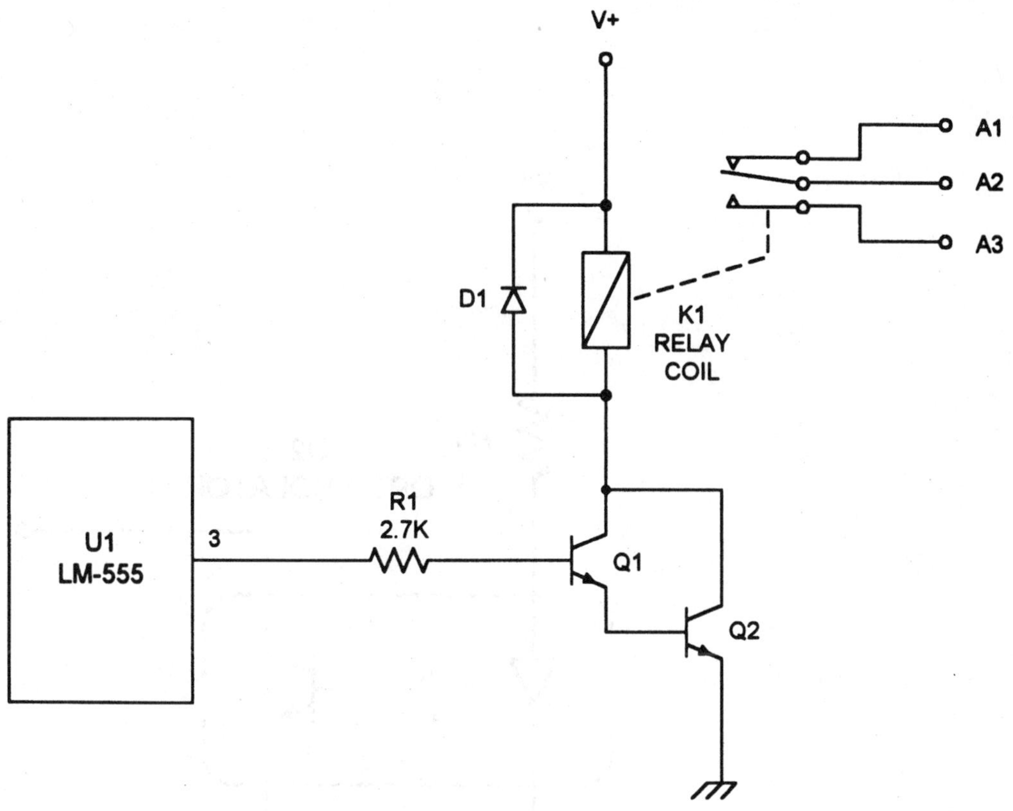

Figure 18-1e. High-gain, high-current relay.

the pair is 100 x 100 = 10,000. Even very large collector currents can be driven with little more than a breath of hot air on the base of Q1...assuming the overall collector current and power dissipation ratings allow it.

Figure 18-2 shows the use of an optoisolator to provide the isolation between the LM-555 output and the load. Although the device shown as U2 has an NPN transistor as the output device, other optoisolators will have silicon controlled rectifiers, triacs, Darlington devices or even resistors. In all cases, however, the output device is light actuated, so will turn on only when the internal light emitting diode is turned on. This conditions occurs when the LM-555 output is LOW.

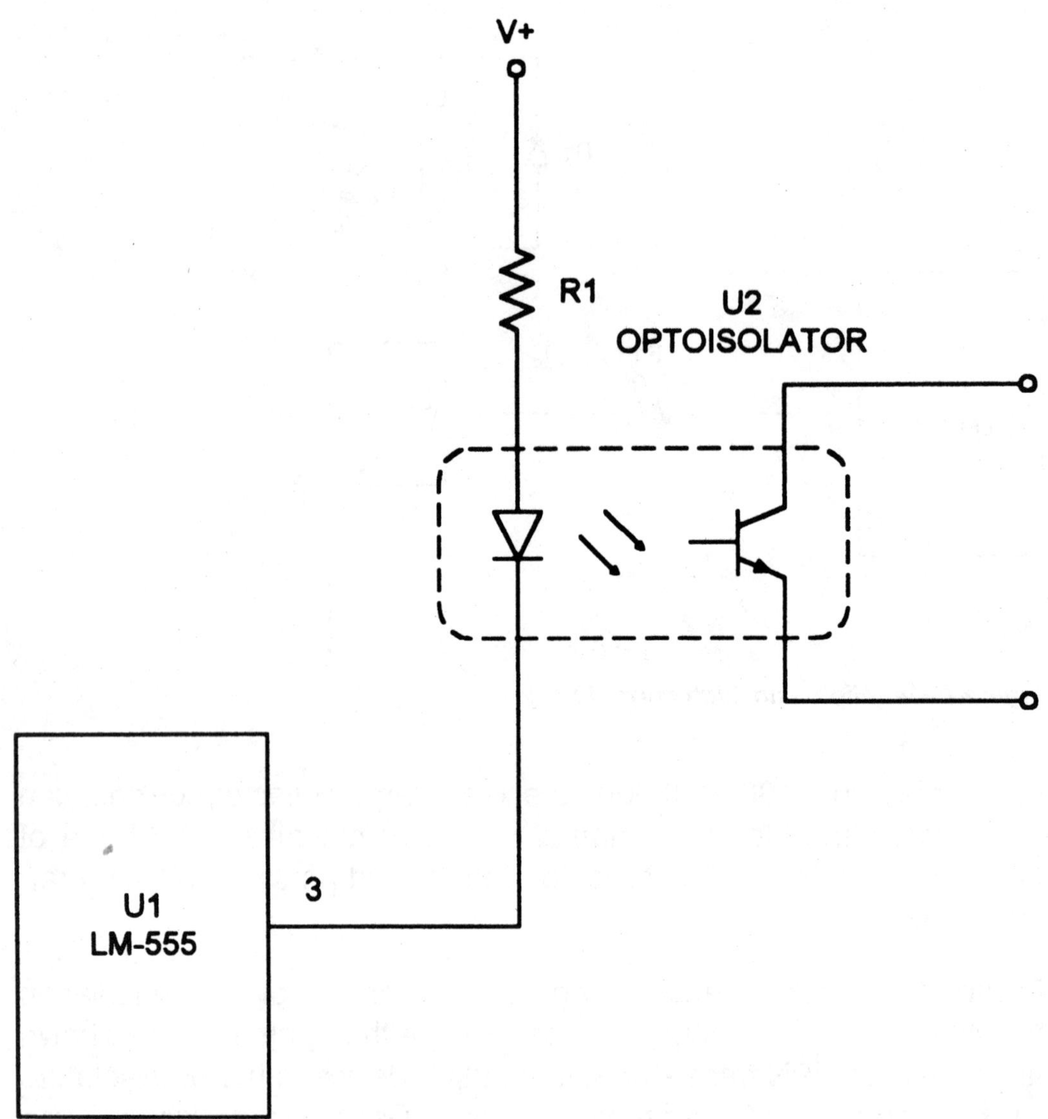

Figure 18-2. Optoisolator interface.

If you want a circuit that actuates when the LM-555 output is HIGH, then connect the LED and resistor between the LM-555 output and ground. The cathode end of the LED (currently connected to pin no. 3) is grounded in that case, while the "hot" end of the resistor is reconnected to pin no. 3.

Chapter 19

Frequency Synchronized Operation

It is often necessary to synchronize an oscillator to some external frequency source. Your television receiver, for example, uses both vertical and horizontal deflection oscillators that are synced to the television broadcast transmitter via sync pulses added to the signal at the station. These pulses are added to assure that the television receiver scans the screen in step with the TV cameras at the station.

You can synchronize the LM-555 astable multivibrator by using a circuit such as **Figure 19-1**. This circuit uses the coincidence of two pulses to trigger the LM-555 astable circuit. The gate shown is a TTL NAND gate (type 7400), although other forms of two-input NAND gate can also be used.

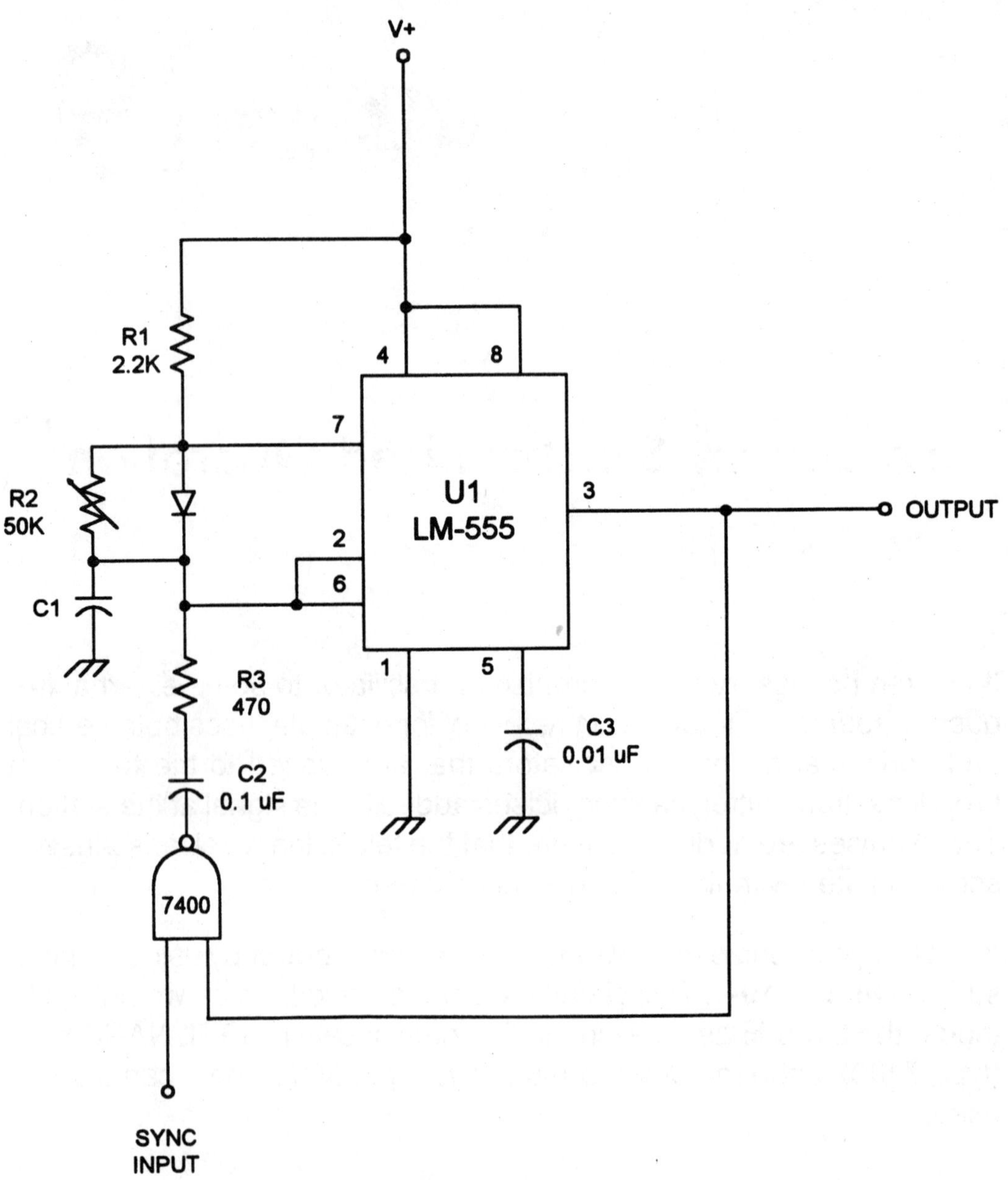

Figure 19-1. Synchronized astable operation.

A NAND gate will produce a HIGH output if either input is LOW, and produces a LOW output only if both inputs are HIGH. As long as the LM-555 output (pin no. 3) is HIGH, then pulses from the synchronization clock (F_S) can pass through the gate to the R-C differentiating network (C2/R3) that triggers the LM-555 device. If the sync-clock frequency disappears, then the astable will free-run at its natural frequency. The operating frequency of the LM-555 and the sync pulses must be either close to each other, or have a harmonic relationship.

Chapter 20

Code Practice Oscillator

Learning the International Morse Radiotelegraph code, aka Morse code, is still necessary for some classes of amateur radio license. It requires practice to send and receive code well. Before one can transmit on the air, it is necessary to practice off-the-air...and that's what this project provides.

One of the tools needed to properly learn the code is a code practice oscillator (CPO). The CPO is an audio oscillator set to a frequency between 300 and 1200 Hz (depending on what is comfortable for you), that can be keyed on and off by a telegraph key.

An example of a code practice oscillator circuit is shown in **Figure 20-1**. The oscillator is an LM-555 astable multivibrator with component values selected to produce about 600 Hz. Component values can be changed to suit your preferences.

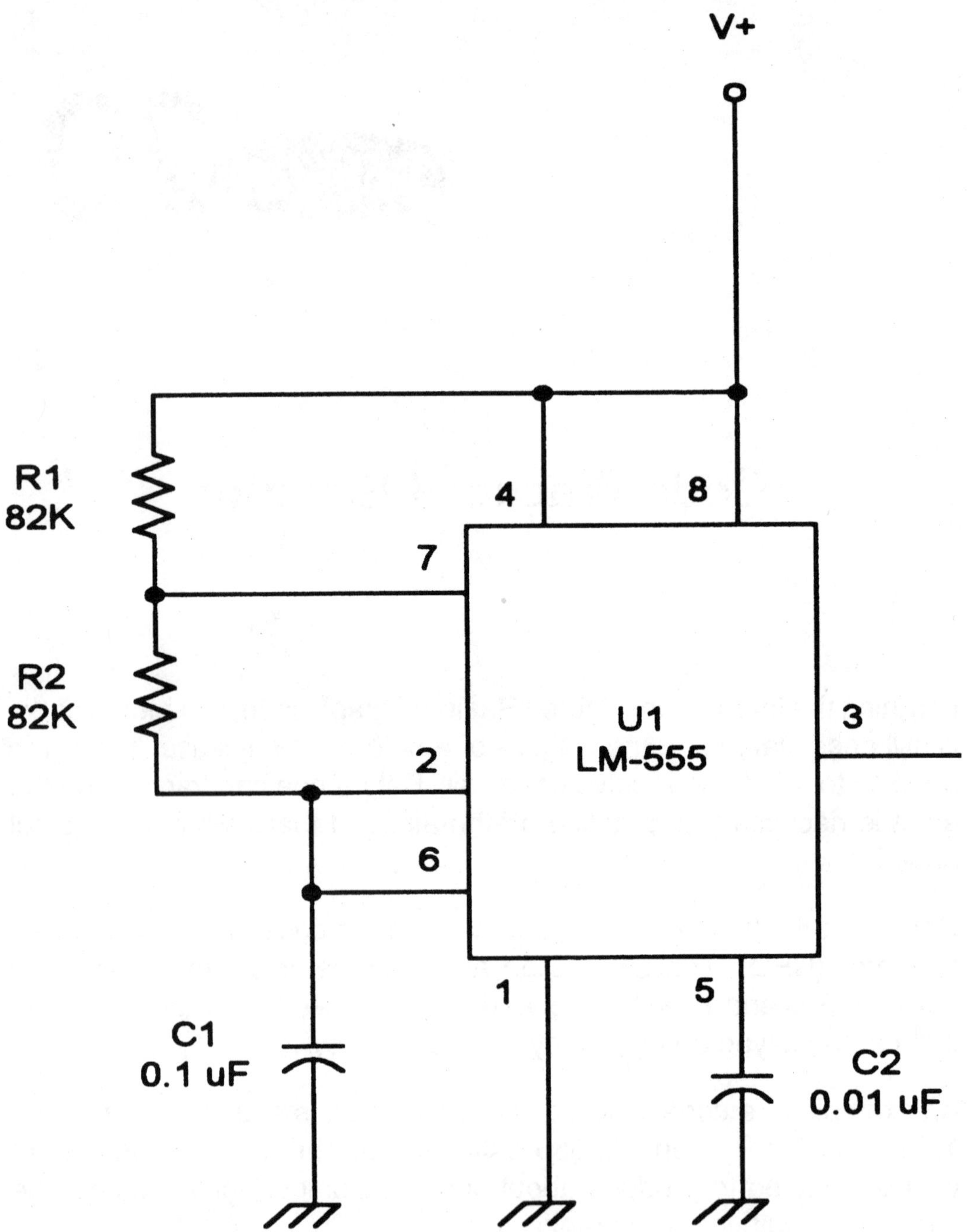

Figure 20-1. Earphone tone generator.

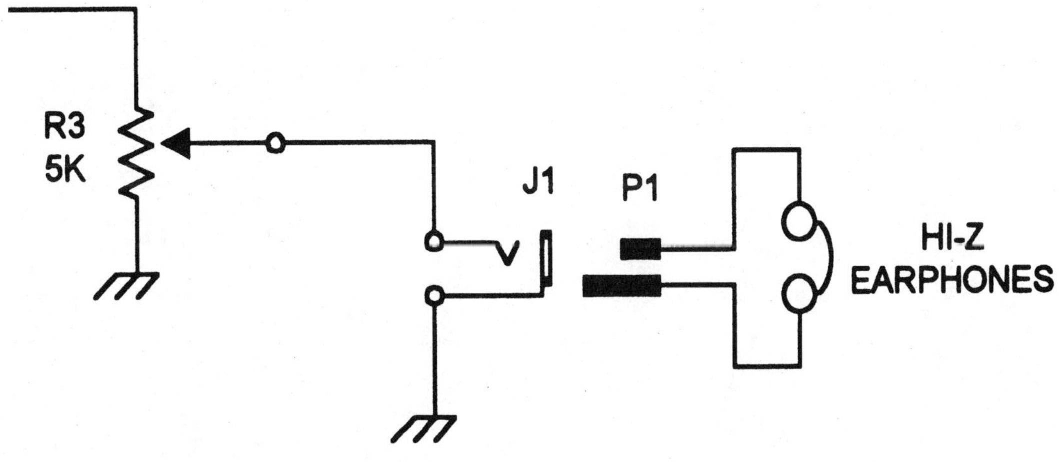
R3
5K
J1
P1
HI-Z
EARPHONES

The LM-555 is capable of driving a small loudspeaker or set of high impedance earphones. The earphones are connected to a 5,000-ohm volume control (R3) and to the series-wired telegraph key (J1). The volume control can be adjusted to produce a comfortable level in the earphones. The operator can turn the tone on and off with a telegraph key.

Chapter 21

Tone Generator

A tone generator is used to produce a nearly sinusoidal audio tone for purposes of testing, signaling and so forth. The LM-555 astable multivibrator will produce the audio tone, but it is in the form of a square wave. **Figure 21-1** shows how to arrange the output circuit for an astable multivibrator to filter out the harmonics of the square wave and thereby produce a sinusoidal (or nearly so) output waveform.

The LM-555 astable section of the circuit produces an output signal consisting of square waves with a frequency set by timing resistors R1 and R2, and timing capacitor C1. The output (pin no. 3) is applied to the input of a two-stage RC integrator. The function of an integrator, in this case, is low-pass filtering of the square wave. The components of this network are selected to have a cut-off frequency that is slightly higher than the fundamental frequency of the LM-555 astable multivibrator, but not close to the second harmonic. This will cut off most of the harmonics and severely attenuate the lower har-

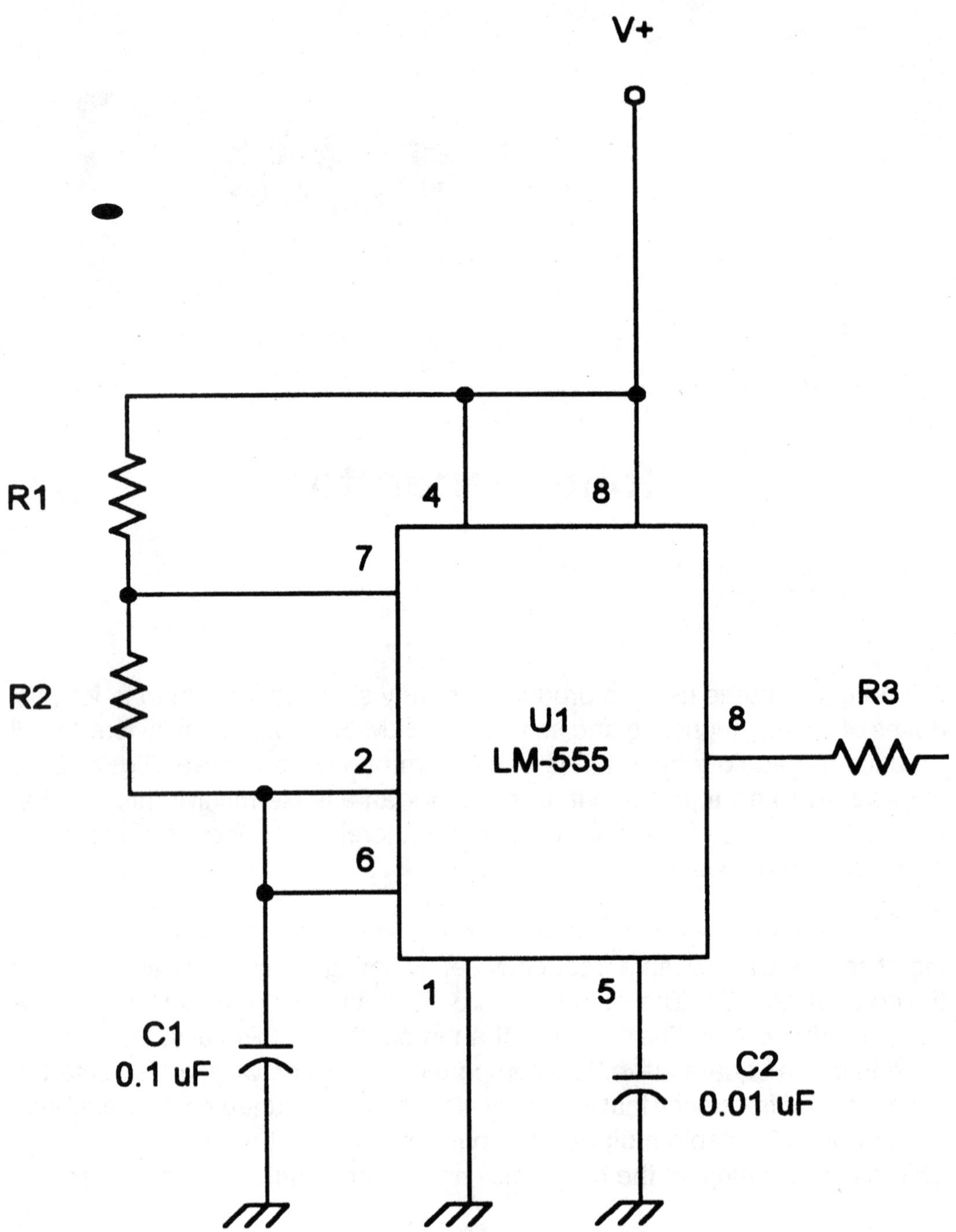

Figure 21-1. Level controlled output circuit.

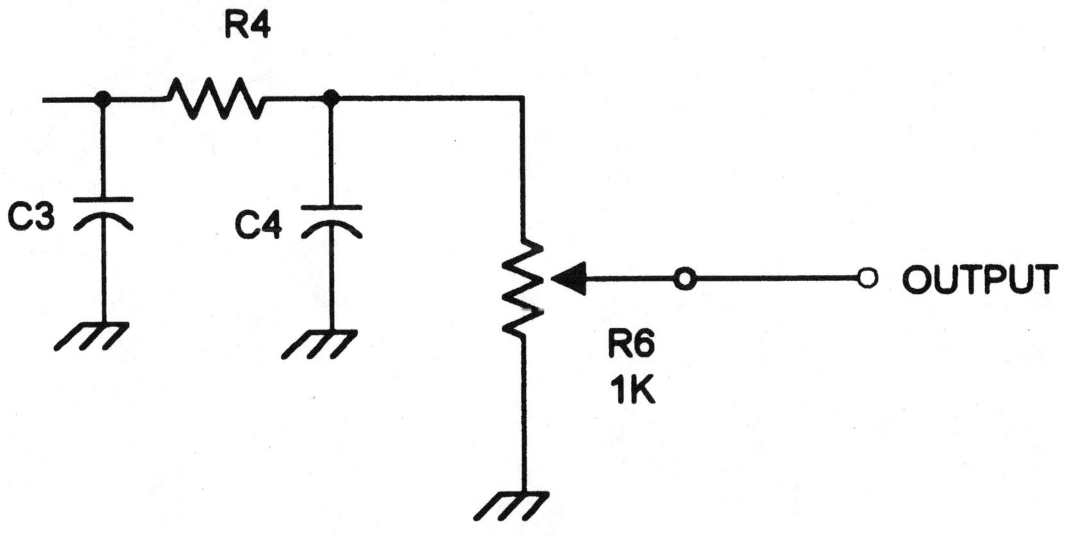
R4
C3
C4
OUTPUT
R6
1K

monics. Additional sections of filtering will provide a more nearly sinusoidal output, but will cost in amplitude. If the amplitude is too low, then an operational amplifier can be used to amplify the output. The component values of the integrator can be selected to:

- $$F = \frac{1}{2\pi R C}$$ *eq. (21-1)*

Where:

F is the operating frequency (Hz)

R is resistance in ohms (Ω)

C is the capacitance in microfarads (µF)

A volume or level control is shown in this circuit, but it is optional. Whether you use it will depend upon your own application. Resistor R5 is used as a voltage divider to produce low-level outputs, and (again) it is optional depending application.

Chapter 22

Warble Siren/Alarm

Have you ever heard those electronic sirens that the police and fire fighters are using today? They don't make the continuous wailing sound of earlier sirens, but sound a lot more like Europeans sirens. They warble between two tones, a high tone and a low tone (beeeeeebooooo beeeeeebooooo). The circuit in **Figure 22-1** is a bitonal audio generator that can be used to form a warble siren.

It can also be used for a unique alarm sound. I once worked in medical electronics at a hospital. One of our engineers designed a respirator alarm using a standard alarm sounding device (2,300 Hz). Unfortunately, so did everything else in the Intensive Care Unit (ICU), even relatively benign alarms. That little human factors fiasco meant that nurses and physicians tended to be desensitized to deadly alarms. The key is to make the important alarms all sound different from the others and from each other. This circuit can help that problem.

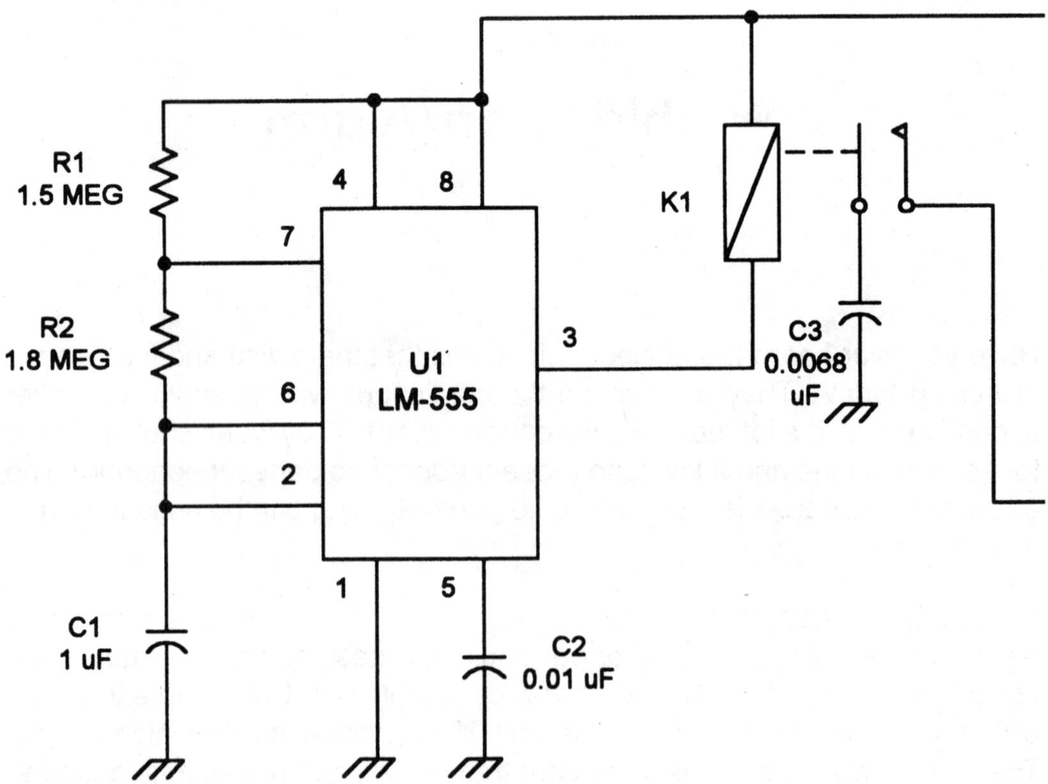

Figure 22-1. Warbling siren.

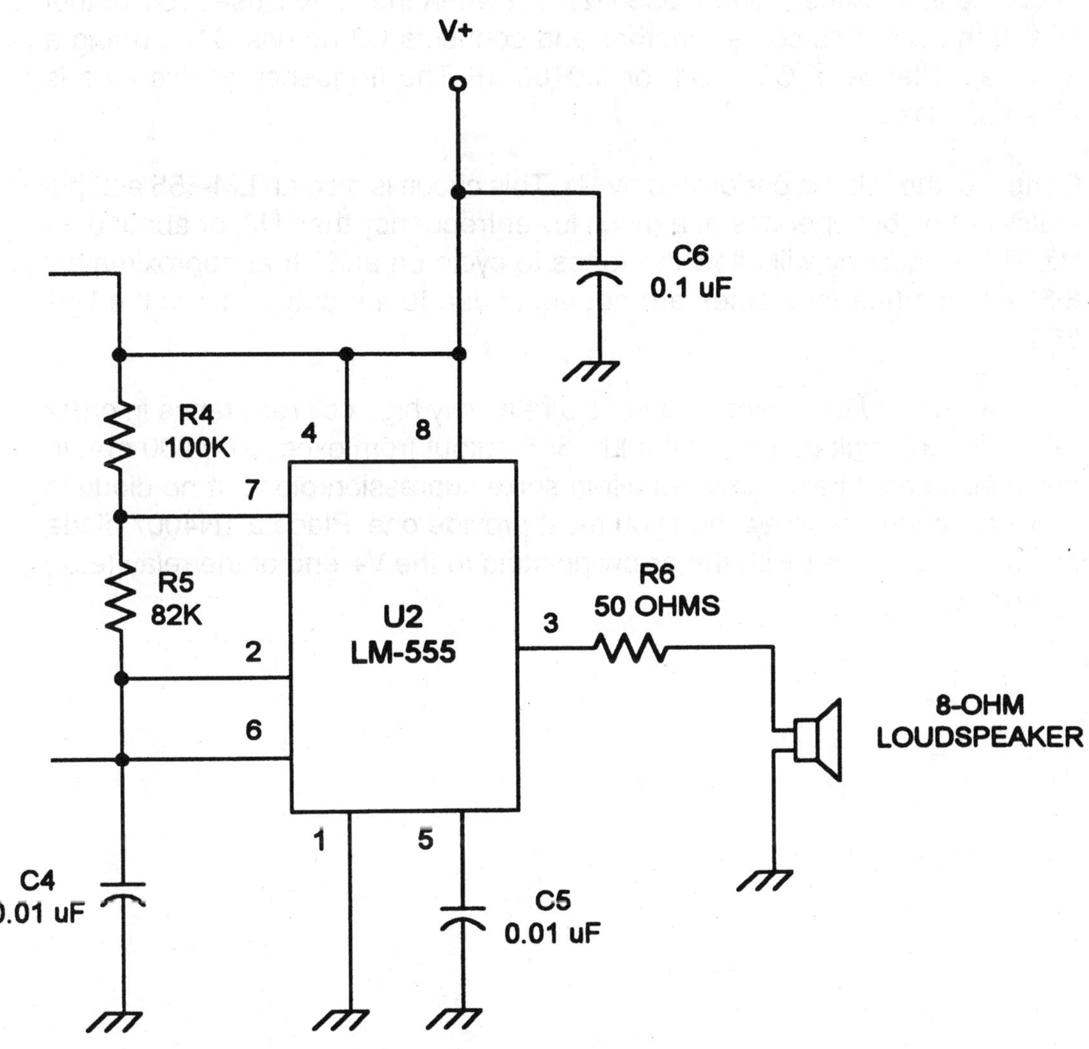
V+
C6
0.1 uF
R4
100K
4
8
7
R5
82K
U2
LM-555
3
R6
50 OHMS
2
6
8-OHM
LOUDSPEAKER
1
5
C4
0.01 uF
C5
0.01 uF

The main siren section is formed by U2, an LM-555 astable multivibrator that will oscillate at either of two frequencies. Two timing capacitors are used in this circuit in order to produce the two different frequencies. When the relay contacts are open (U1 output HIGH), then only C4 is in the circuit and the high tone is produced (about 550 Hz). But when the relay closes (U1 output LOW), the contacts come together and connects C3 across C4, forming a total capacitance of C3 + C4, or 0.0168 µF. The frequency at this time is about 350 Hz.

Control of the relay is performed by U1. This circuit is also an LM-555 astable multivibrator, but operates at a much lower frequency than U2, or about 0.28 Hz. This frequency will allow the tones to cycle on and off at approximately 3.5 seconds (the tone times are not equal due to the duty cycle of the LM-555).

Relay K1 should be selected to have a relatively high coil resistance in order to prevent the sink current of the LM-555 output from exceeding 200 mA. In addition, it should also have a built-in spike supression diode. If no diode is provided inside the relay, then you must provide one. Place a 1N4007 diode across the coil of K1 with the arrow pointing to the V+ end of the relay (e.g., cathode to V+).

Chapter 23

Two-Phase Digital Clock

The project shown in **Figure 23-1a** is a two-phase digital clock for use with certain types of digital circuits, especially microprocessors. A two-phase clock must produce clock pulses on two separate lines that are non-coincident with each other. In addition, there is often a need to ensure that the settling time of circuits being driven is exceeded before a new clock pulse is received. This becomes especially important when interfacing devices such as analog-to-digital converters (ADC) or digital-to-analog converters (DACs), or when using ADCs and DACs with sample-&-hold devices.

Figure 23-1a shows a two-phase clock circuit that accomplishes the non-coincident requirement by driving the NOR gates with the output of a J-K flip-flop. We might be tempted to use the flip-flop without the NOR gates. After all, the two outputs of the J-K flip-flop are complementary, so one will be HIGH when the other is LOW. Although this situation fulfills our requirement of non-coincidence, it does not allow for the settling time of the clocked circuits. In order to accomplish the latter goal we must synchronize the system with the NOR gates.

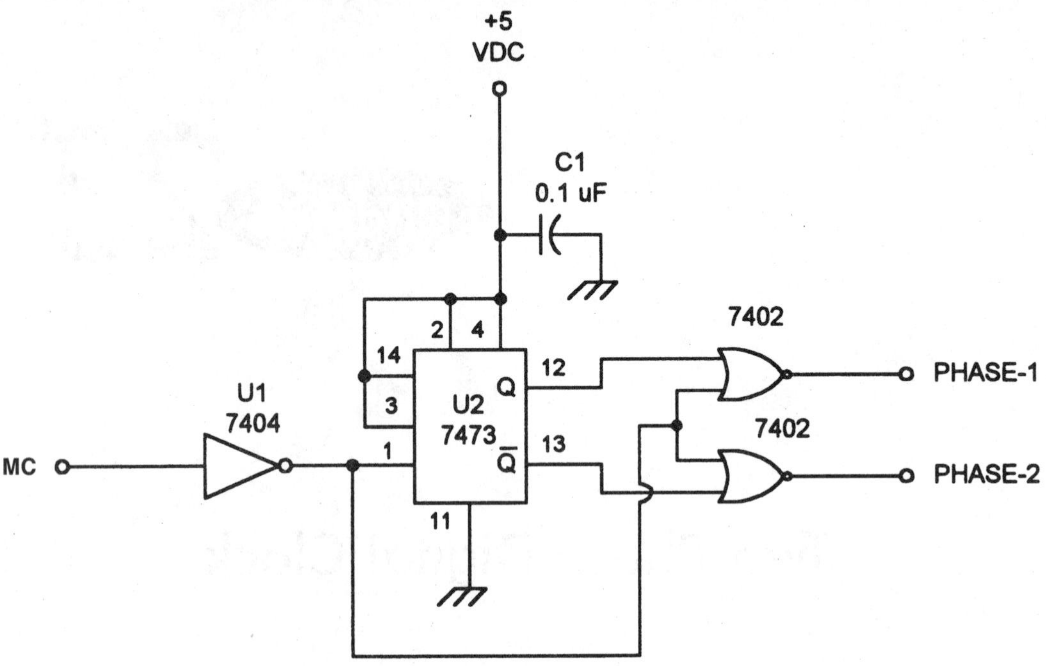

Figure 23-1a. Two-phase clock.

The basic rules for the NOR gate are:

1) A HIGH on either input produces a LOW output, and
2) Both inputs must be LOW for the output to be HIGH

With these rules in mind, let's figure out how the circuit of **Figure 23-1a** works. Note that the master clock (MC) signal, applied to the input of the inverter stage (U1), is inverted before being applied to the input of the J-K flip-flop. The inverted MC signal is applied to one input of both NOR gates. The other input of each NOR gate is connected to one output of the J-K flip-flop. We find that the Q output is connected to the gate that controls phase-1, while the NOT-Q output of the J-K flip-flop controls the phase-2 output of the two-phase clock. The rules for the NOR gate tell us that it is necessary for both inputs of either NOR gate to be LOW before an output pulse can occur.

Let's consider phase-1. The two inputs of the phase-1 NOR gate are the inverted master clock signal and the Q output of the J-K flip-flop. When the first clock pulse arrives at time T1 (**Figure 23-1b**), the inverted MC line drops LOW. This makes the Q output of the flip-flop go HIGH. Since there is a HIGH on one input of the NOR gate, the phase-1 output is LOW. It is not until the second MC pulse arrives that we find a situation in which both inputs of the NOR gate governing phase-1 are LOW. When this pulse comes along, the phase-1 clock pulse is generated. At that same time, the phase-2 clock pulse is inhibited by a HIGH on one input of the phase-2 NOR gate.

This circuit produces output pulses at a frequency of one-half the master clock frequency. The complementary nature of the Q and NOT-Q outputs of the flip-flop ensure that only one pulse is generated at a time, while the NOR gates ensure that a full clock period will exist between each pulse. This will allow circuits being clocked to settle for the next happening.

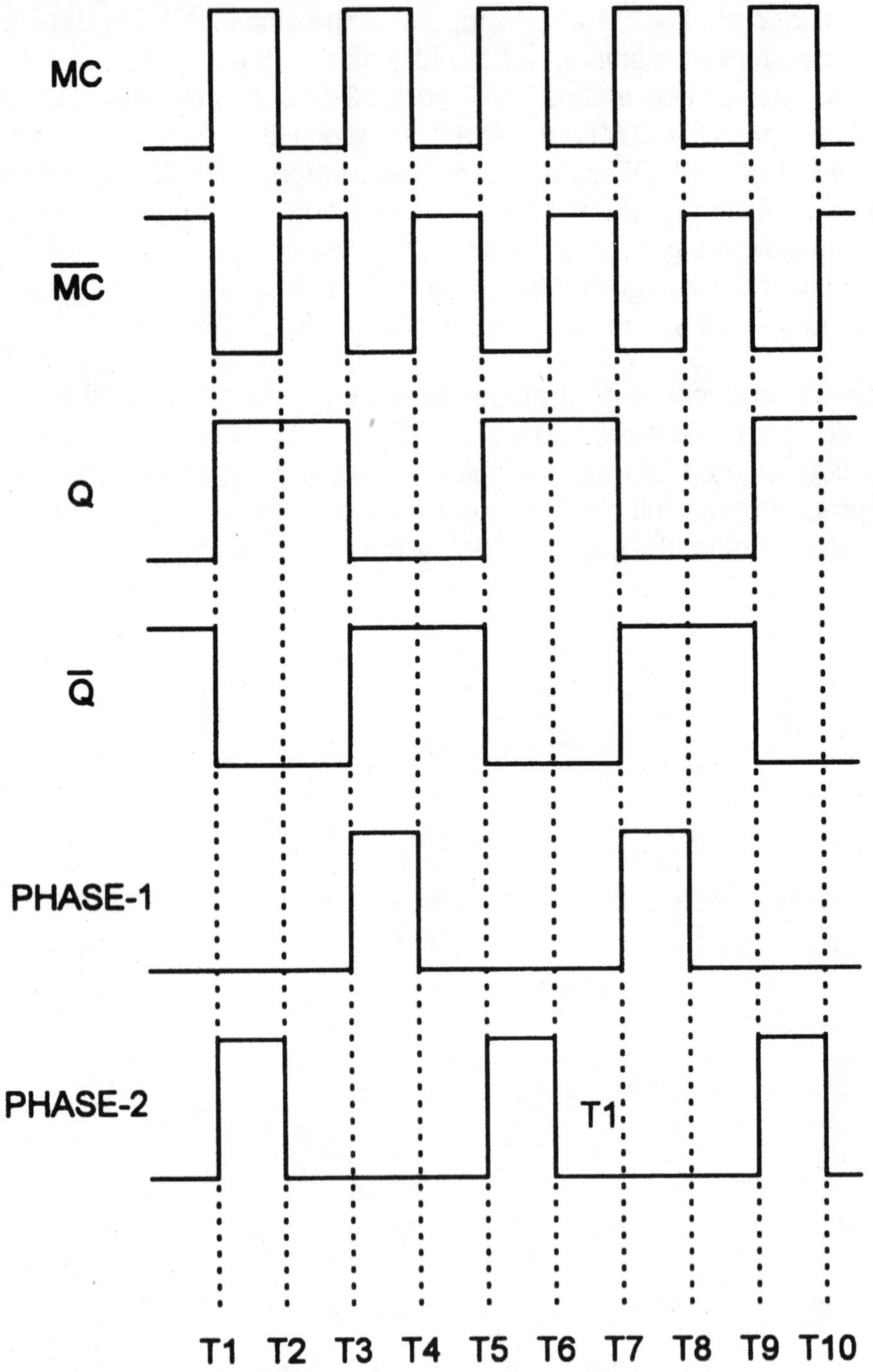

Figure 23-1b. Timing diagram.

Chapter 24

Sixteen-Phase Clock

The digital clock shown in the previous project is capable of driving a large number of different types of digital projects, including microprocessors. But what do you do if you want to create a multiple phase clock? The use of gating schemes can create a large array, and that can lead to errors (not to mention a large bill for +5 volt power supplies). The circuit in **Figure 24-1** allows us to generate up to sixteen clock phases with just two integrated circuits.

The heart of the circuit is a 1-of-16 data distributor chip. This 24-pin DIP IC examines a four-bit binary word input and decodes the word to a unique output condition (i.e., in which one output is LOW and all others are HIGH). The binary word on the four inputs lined to the 74154 in **Figure 24-1** will cause only one output at a time to be unique. If switch S1 is open, then the selected output will be HIGH and all others are LOW. But if S1 is closed, then pin no. 18 is LOW so the selected output will be LOW and all others are HIGH. We can call switch S1 an output polarity select switch. The table below gives the codes for the selections:

0000	1
0001	2
0010	3
0011	4
0100	5
0101	6
0110	7
0111	8
1000	9
1001	10
1010	11
1011	12
1100	13
1101	14
1110	15
1111	16

There are no settling states between phases of this circuit (as in the previous circuit). The next phase becomes active immediately after the previous phase goes inactive.

The input word for the 74154 is formed by a four-bit binary counter, the TTL type 7493. This counter will continue to increment through the range until power is turned off because it is connected with the reset terminals (2 and 3) to ground.

We may select phase lengths considerably less than sixteen using this same circuit. What is required is some feedback to reset the counter to 0000 when the desired count is reached. The reset pins of the 7493 want to be LOW for

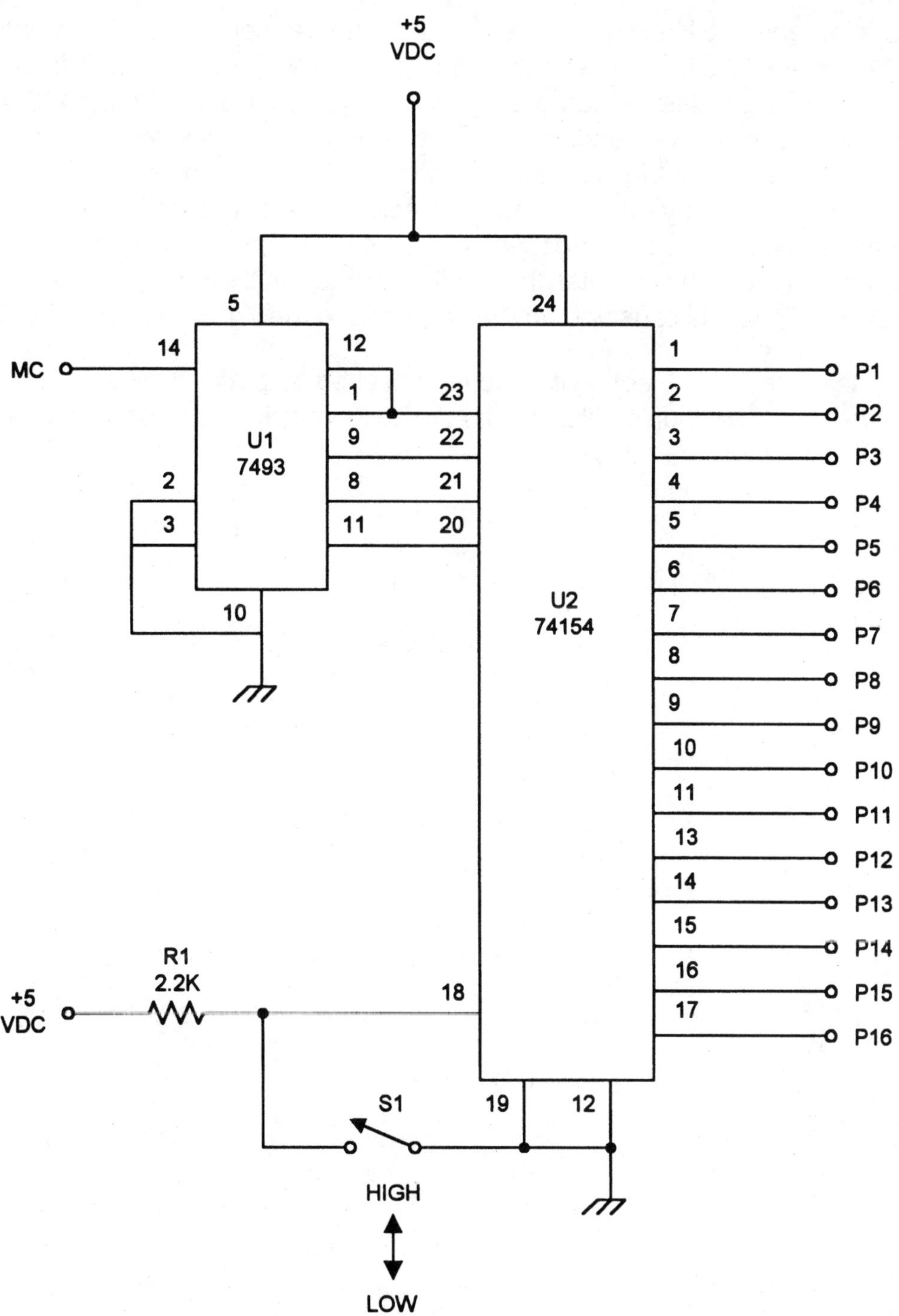

Figure 24-1. Sixteen-phase clock.

operation and HIGH for reset. We could, therefore, select the switch option that gives a HIGH for the desired output and LOWs for all others. We could then connect the desired terminal output to the reset pins of the 7493. For example, suppose we need to make a twelve-phase clock. We would then connect phase-12 of **Figure 24-1** (i.e., pin no. 13) back to pins 2 and 3 of the 7493. As long as any other output is selected, the reset pins of the counter remain LOW, so the count proceeds in the usual manner. But when the clock has incremented the counter the twelth time, a HIGH is sent to the 7493 reset lines, and this causes both the counter and the 74154 to reset to 0000.

If we select the switch option in which the desired output is LOW and all other outputs are HIGH, then it will be necessary to feedback the reset pulse through an inverter stage.

Chapter 25

Power Failure Alarm

There are times when it is imperative to know when the AC power mains have failed. We might, for example, be using AC power to run an electrical heater in a greenhouse or fish pool, or some other place that must remain warm. If the mains power fails, many plants or fish could be lost. We can use the circuit of **Figure 25-1** to warn of a power failure.

This circuit is an LM-555 astable multivibrator connected so that it turns on only when gating transistor Q1 is forward biased. The bias on the base-emitter junction of transistor Q1 is the potential that appears across capacitor C4. This capacitor is charged from two DC sources: the current through R3 and the current through R5. The current through R3 is derived from the V+ DC power supply that operates the LM-555, while the current through R5 is derived from a negative polarity pulsating DC power supply operated from the 120 VAC mains. If the AC mains voltage is present, then the potential across C4 will be close to zero or slightly negative. This situation keeps transistor Q1 reverse-biased and the astable multivibrator is silent. But, if the mains volt-

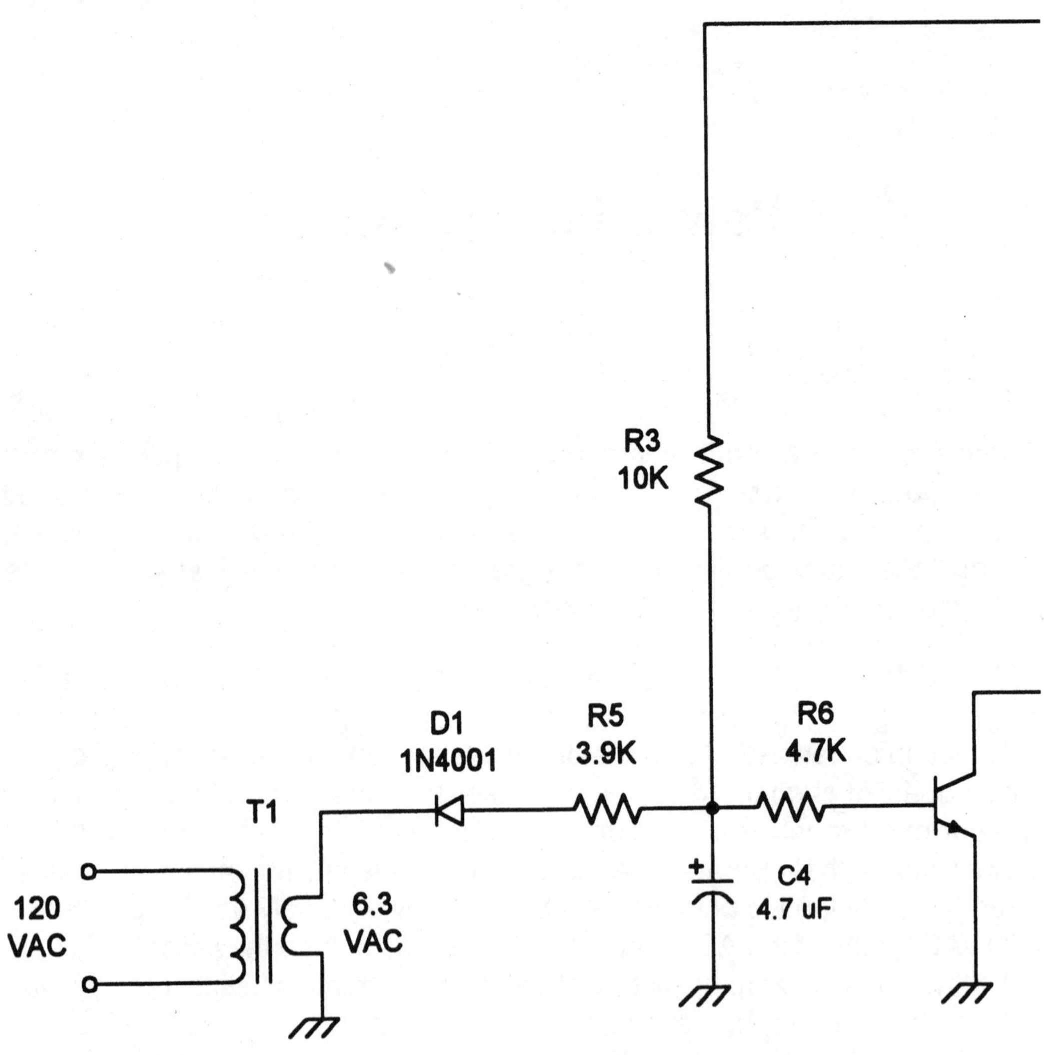

Figure 25-1. Power line monitor.

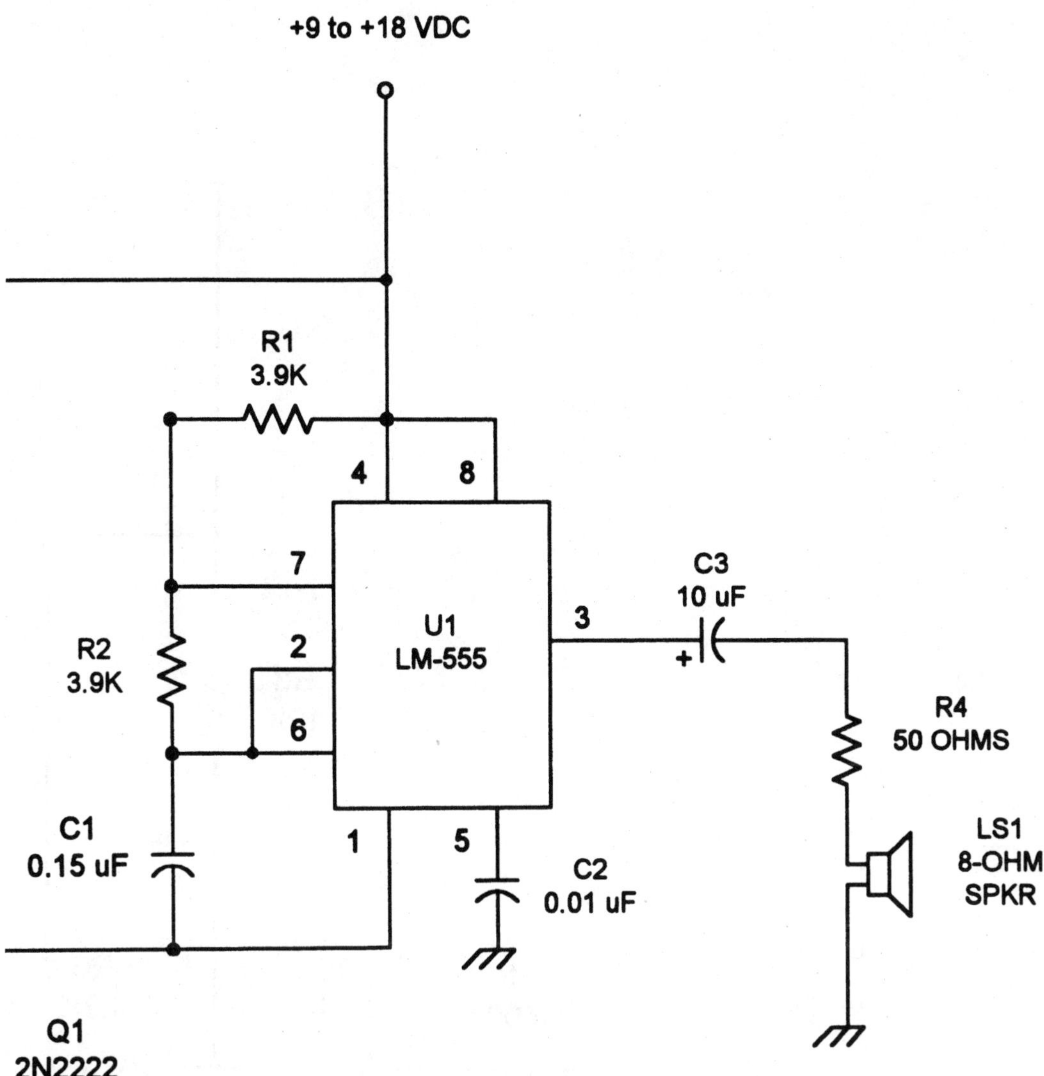
+9 to +18 VDC
R1
3.9K
4
8
7
2
6
1
5
U1
LM-555
3
C3
10 uF
+
R2
3.9K
R4
50 OHMS
C1
0.15 uF
C2
0.01 uF
LS1
8-OHM
SPKR
Q1
2N2222

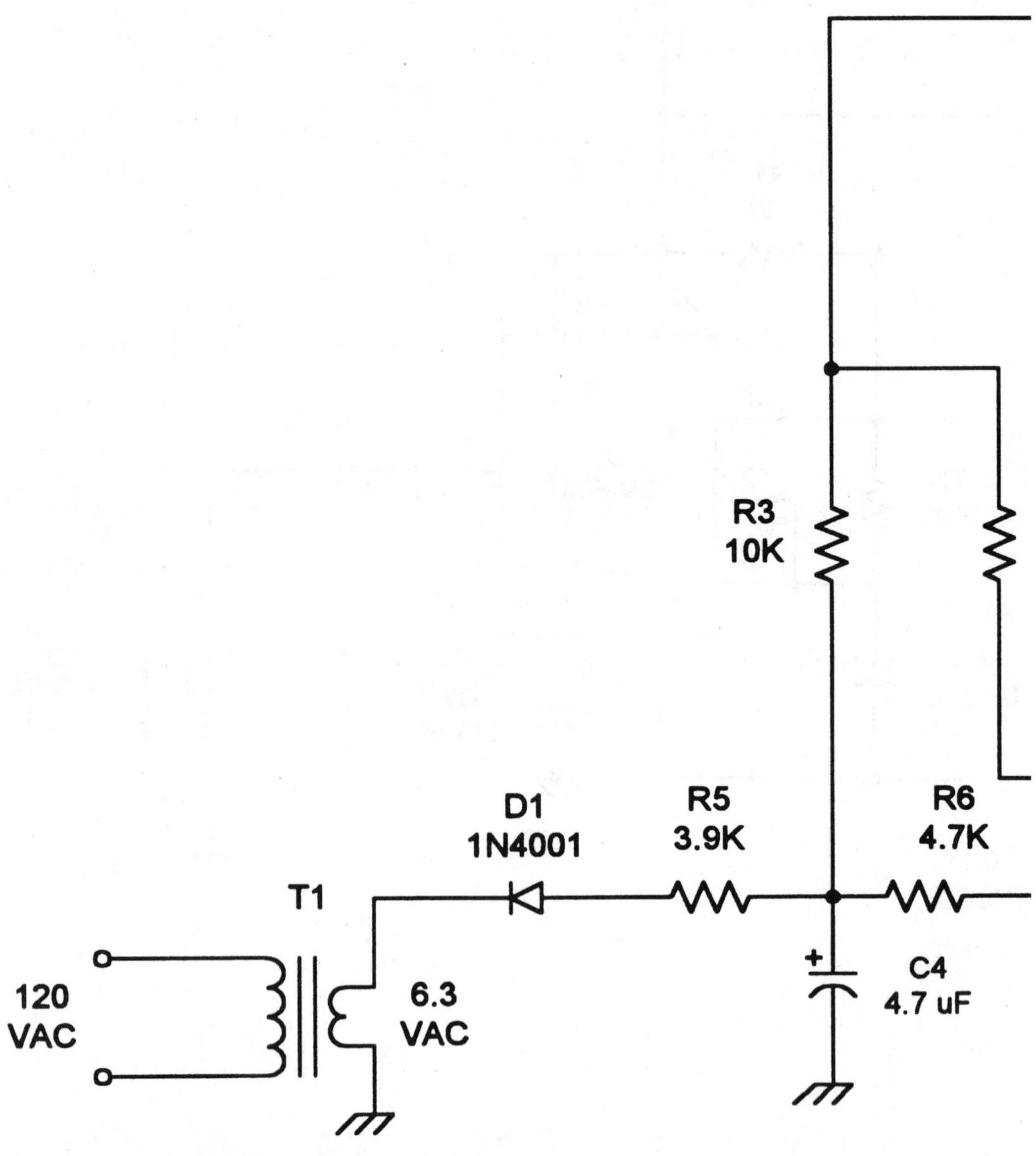

Figure 25-2. Alternate power line monitor (one-shot) circuit.

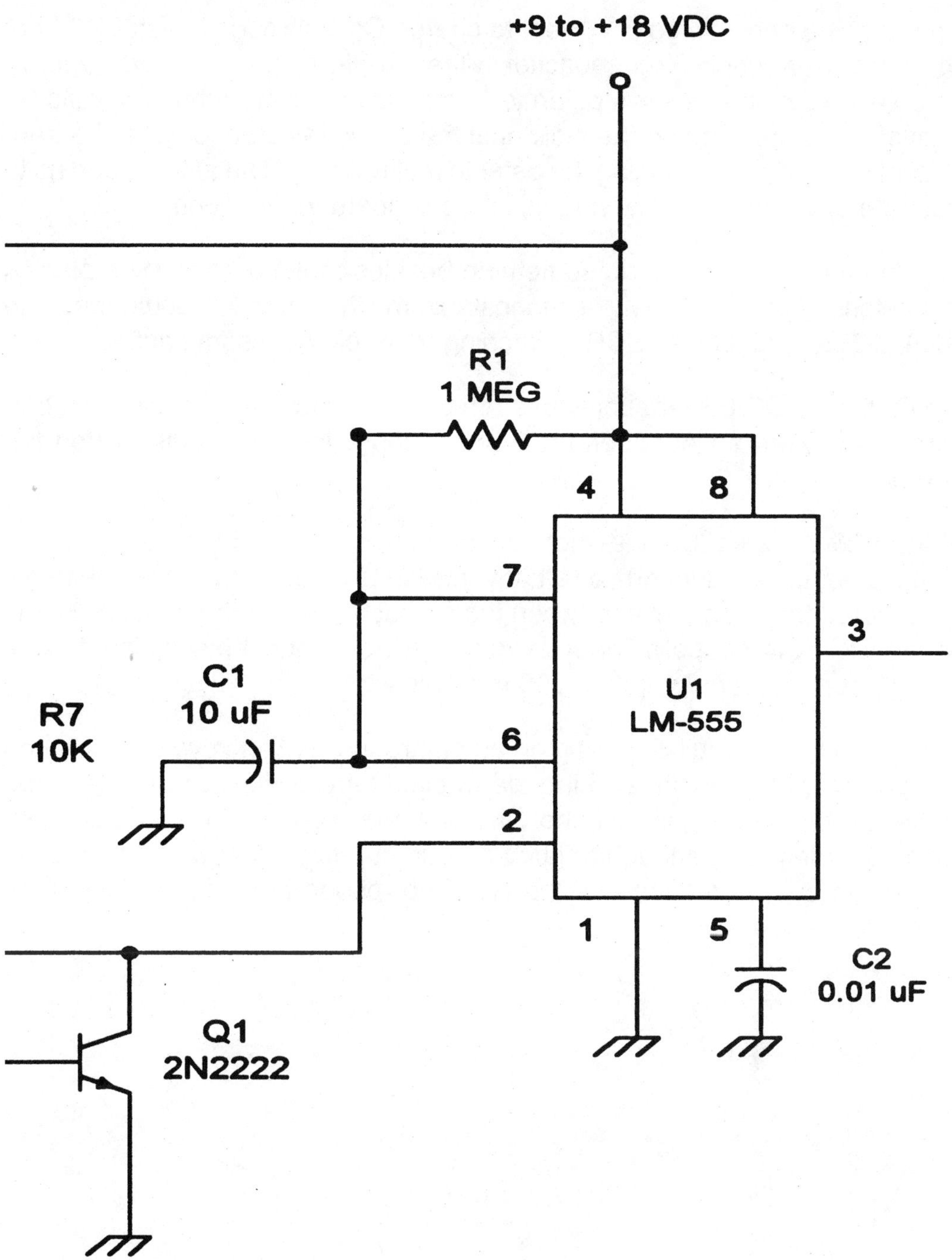
+9 to +18 VDC
R1
1 MEG
4
8
7
3
C1
10 uF
U1
LM-555
R7
10K
6
2
1
5
C2
0.01 uF
Q1
2N2222

age fails, the only source of current to charge C4 is through resistor R3 from the V+ power supply. The capacitor will soon discharge from the negative value and take on a positive polarity. When this polarity reaches 0.7 volts (or a little less depending on the individual transistor selected for Q1), the transistor is turned on, grounding the astable multivibrator. The astable begins to oscillate at about 1,000 Hz, and this tone is heard in the loudspeaker.

We can make this circuit do something besides make a noise by replacing the astable multivibrator with a monostable multivibrator. We could then use the LM-555 to trigger an SCR or latching relay, causing some action.

Note: the V+ DC power supply used to power the LM-555 should be a battery. If it is from an AC operated power supply, then it too dies when the power goes off.

Figure 25-2 shows the one-shot version of this circuit. The output circuit is deleted because it depends on what you wish to control. If you want to use a small relay, then connect it between the output of the LM-555 (pin no. 3) and the V+ DC power supply. The relay coil resistance must be sufficient to limit the DC current to the output to 200 mA or less.

One use for this circuit is turning on an emergency lighting system when the AC power fails. The light and the alarm circuit are powered from the same battery. If the power goes off, then the one-shot turns on for about ten seconds, certainly long enough to latch an output relay. If you wish to increase the length of time required for the circuit to respond to a power loss, then increase the value of C4.

Chapter 26

DC-to-DC Converter

Figure 26-1 shows the circuit for a small negative DC power supply project. It can be used to supply small amounts of power to operational amplifiers, and other circuits that require dual polarity DC power supplies. It may be that the main portions of the circuit are monopolar devices, or there is a need to keep the expense and weight down by using only one battery. This circuit will allow you to use a battery for the +12 VDC supply and generate the -12 VDC from this circuit.

The power supply shown here operates from a positive DC power supply in the +9 to +18 VDC range. It consists of an LM-555 astable multivibrator driving a voltage doubler rectifier filter. The resistors and capacitor in the timing network (C1, R1, R2) are selected to provide an operating frequency of about 10 Hz.

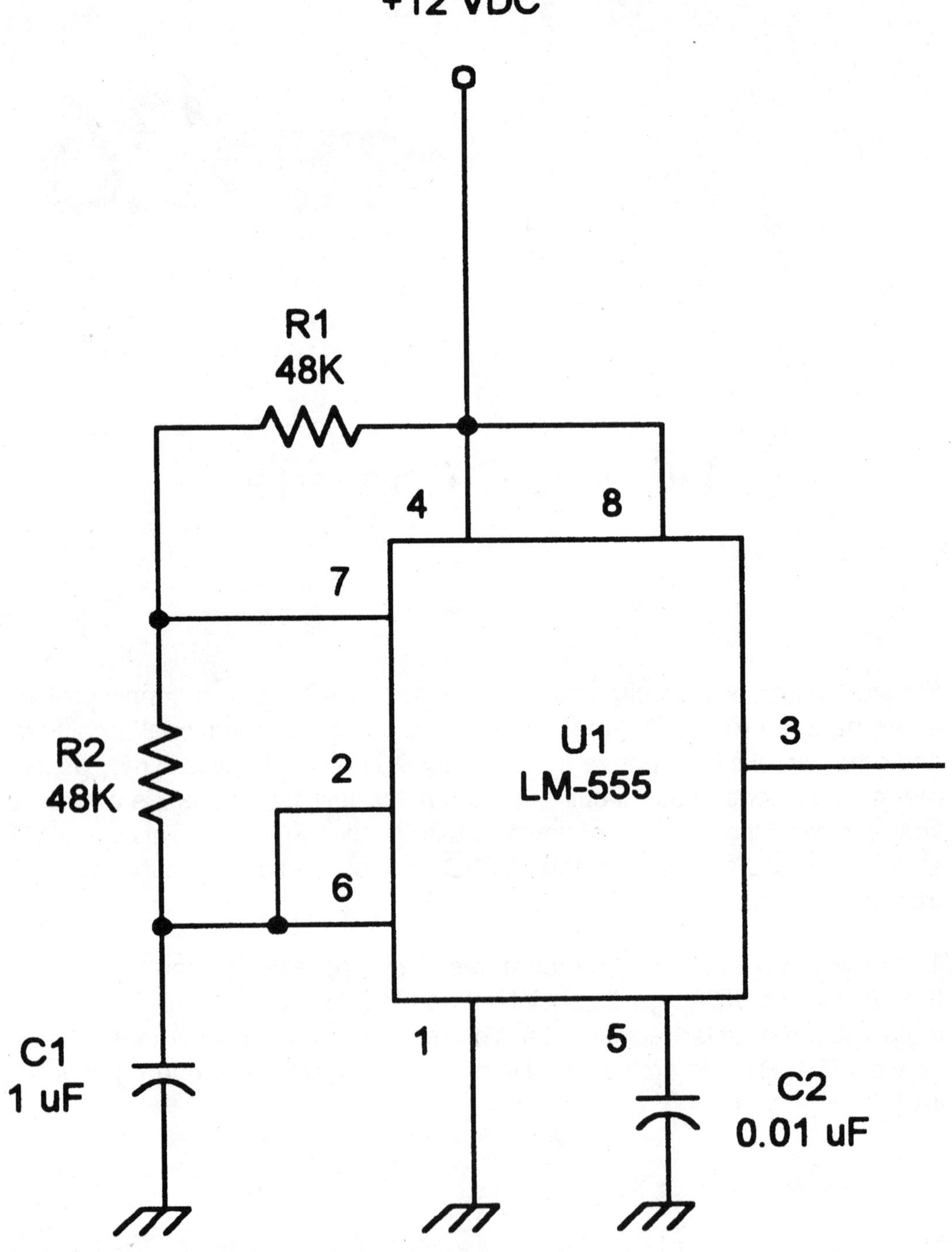

Figure 26-1. Negative voltage output power supply.

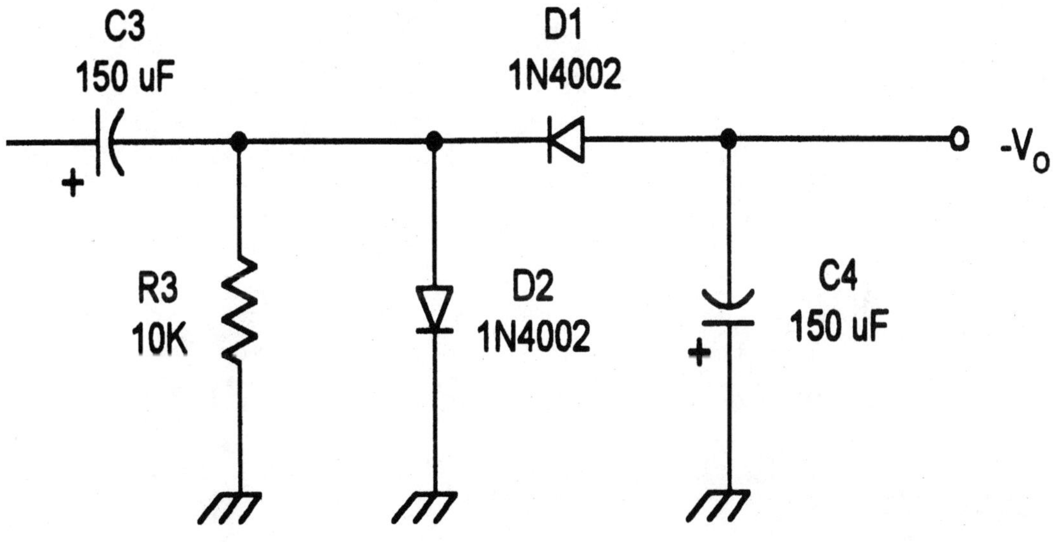
C3
150 uF
+
D1
1N4002
-V$_O$
R3
10K
D2
1N4002
C4
150 uF
+

The output of the LM-555 astable multivibrator is connected to a half-wave voltage doubler circuit consisting of diodes D1 and D2, plus capacitors C3 and C4. The diodes are connected to produce a negative output voltage (reversing both would produce a positive output). This voltage will not be exactly double the supply voltage, even though the circuit is a voltage doubler, because of the duty factor of the output signal. The total negative voltage magnitude will be nearer the DC supply voltage magnitude (but of opposite sign). If this circuit is followed by a negative output three-terminal voltage regulator (e.g., 7912, LM-320-12, etc.), then the output voltage will be -12 VDC and stable. As shown, the voltage output is unregulated.

Chapter 27

Visual Continuity Tester

Continuity testers are instruments that will tell you whether or not a circuit is continuous. An ohmmeter can be used as a continuity tester, as can a combination of a battery and a light bulb. The circuit in **Figure 27-1** is special type of continuity tester in that it will latch the reading and hold it until the RESET button is pressed. This circuit will help immensely in the case where a single technician must dope out a large circuit. In one case, I had to dope out a large public address system. There was an electrical panel in the control room where all the wires from the rest of the building were terminated, but we had no way of knowing which wires went to which rooms. Fortunately, all of the wires were paired. It took two technicians, working with walkie-talkies, to dope out the circuit. The circuit shown in **Figure 27-1** will allow one person to check out such a circuit. The probes can be connected at the remote room, and then a trial short circuit placed across the suspected pair. If the light comes on, then the circuit is continuous—and the light will remain on so the technician can come back and check it.

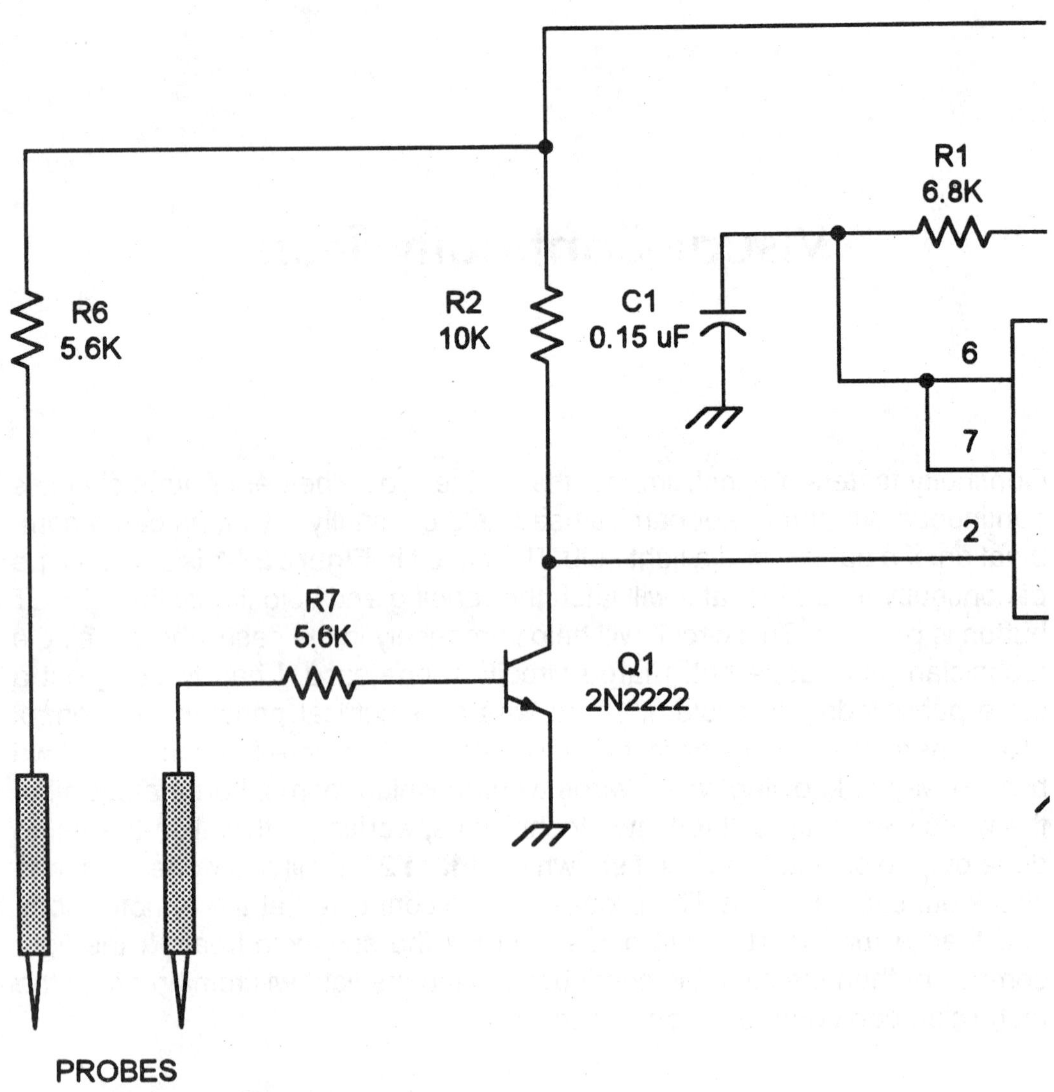

Figure 27-1. Visual continuity tester.

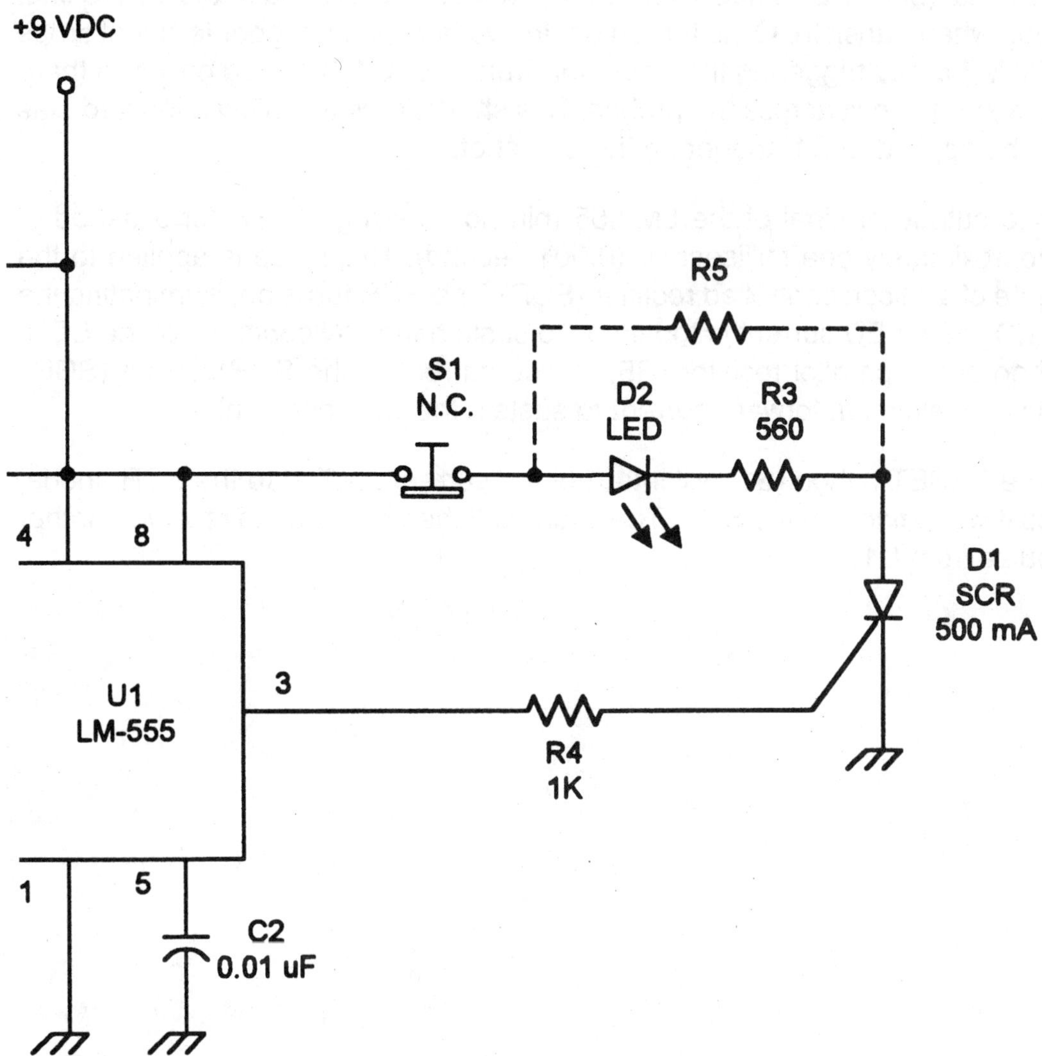
+9 VDC
R5
S1
N.C.
D2
LED
R3
560
4
8
D1
SCR
500 mA
3
U1
LM-555
R4
1K
1
5
C2
0.01 uF

U1 is an LM-555 wired as a monostable multivibrator (one-shot). The trigger terminal (pin no. 2) is held HIGH by resistor R2, connected to the V+ DC line. But, when transistor Q1 is turned on, the voltage at the trigger terminal drops LOW thereby triggering the one-shot. Transistor Q1 is turned on when there is a short circuit across the probes. This short circuit will allow a forward bias to be applied to Q1, triggering the one-shot.

The output terminal of the LM-555 (pin no. 3) will go HIGH for a period of approximately one millisecond (0.001 second). This pulse is applied to the gate of a silicon controlled rectifier (SCR). The SCR turns on, illuminating the LED. If the LED current is too small to sustain the ON condition of the SCR, then add a parallel resistor (R5) to add current to the SCR current (SCRs have a minimum forward current to sustain the ON condition).

The RESET button (S1) will interrupt the current applied to the SCR anode, so it will extinquish the SCR. The circuit will then be ready to receive another pulse from U1.

Chapter 28

Aural Continuity Tester

Type I

Figure 28-1 shows the circuit of a special type of continuity tester. This tester allows the operator to hear when the circuit is continuous. The use of a visual continuity tester or an ohmmeter requires us to look at the tester in order to see if the circuit is continuous. But this tester allows you to listen for the indication of continuity. This type of indication is especially useful when working on a large switch box, chock full of wires and thousands of little terminals. Take your eye off the work for just a second, and you lose your place.

This circuit is an LM-555 astable multivibrator operating at a frequency of 1,000 Hz. The output terminal is connected to a loudspeaker. Note that resistor R3 is optional and is used to reduce the volume of the LM-555 astable multivibrator. The value will depend on the impedance of the loudspeaker and the desired volume level. The probes are used to supply V+ DC voltage to the LM-555 device. If the circuit under test is continuous, then the astable multivibrator will receive V+ and therefore turn on.

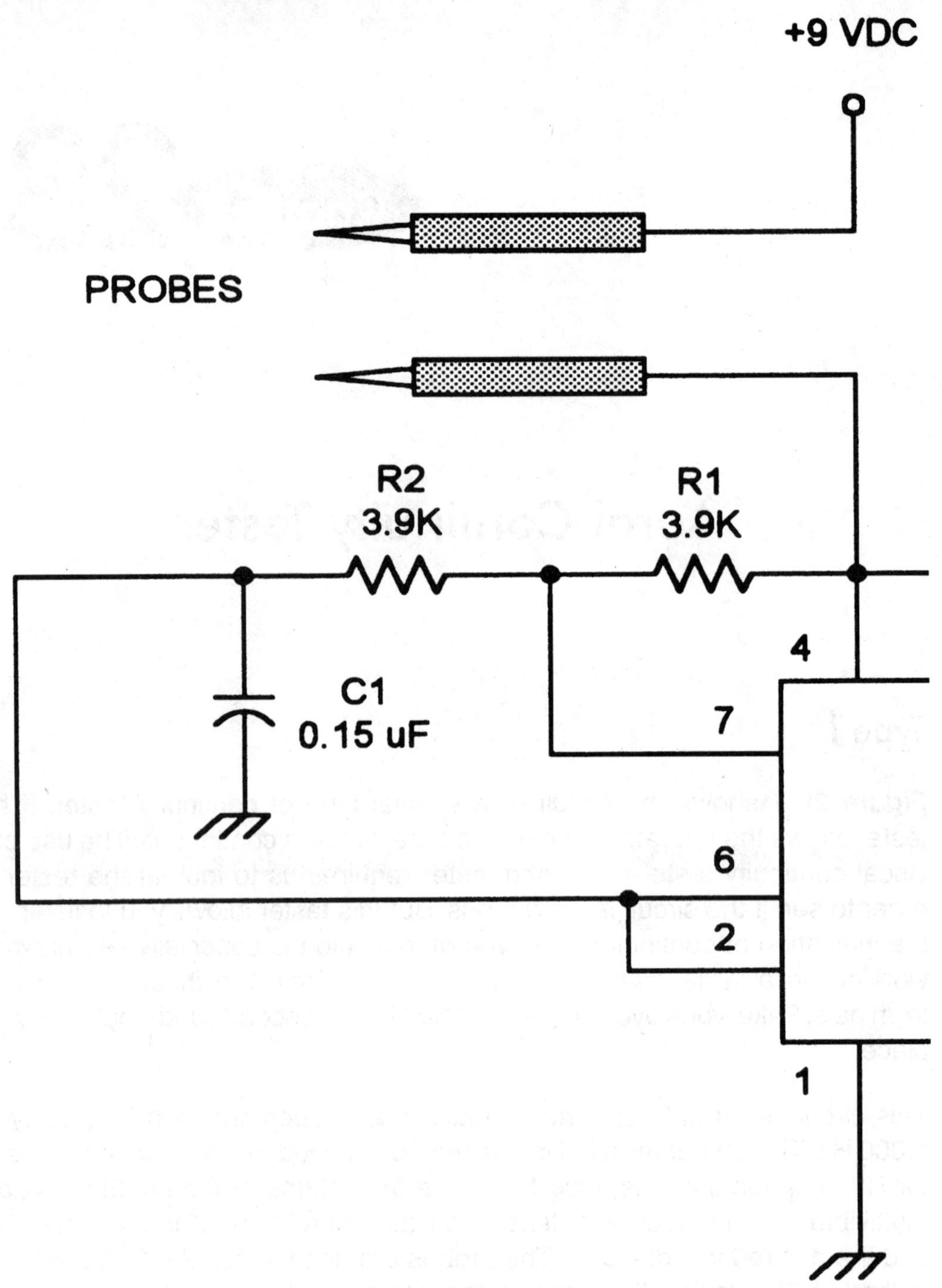

Figure 28-1. Aural continuity tester — Type I.

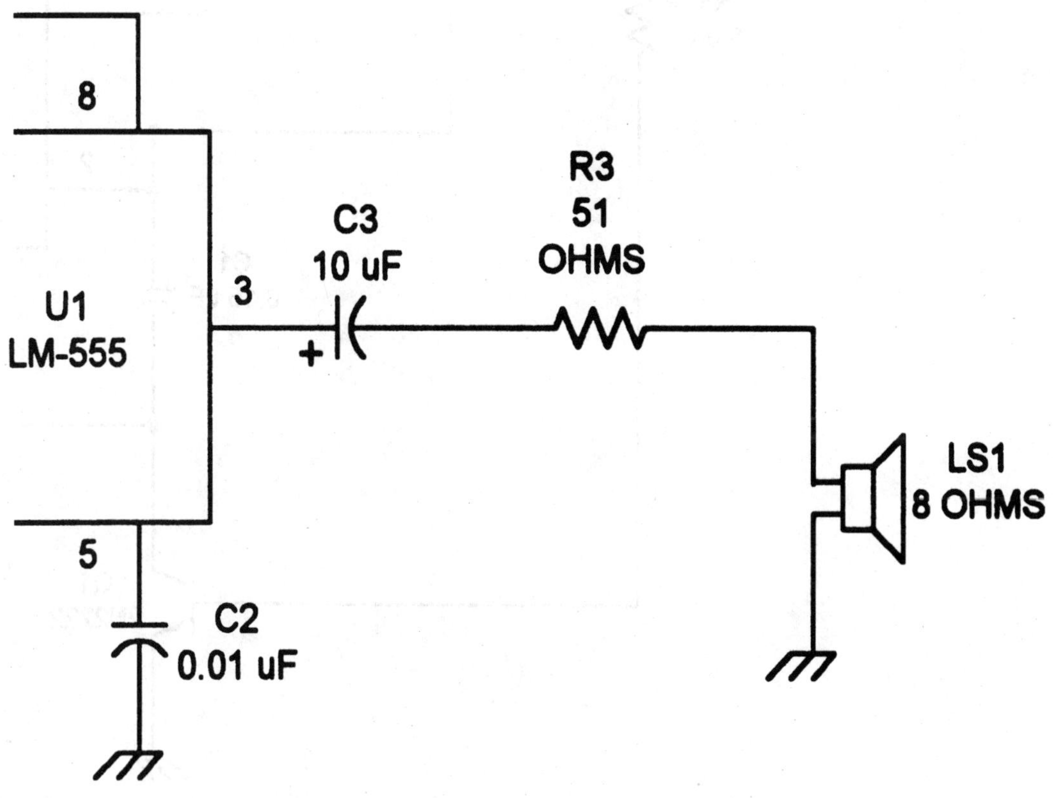
8
U1
LM-555
3
C3
10 uF
+
R3
51
OHMS
LS1
8 OHMS
5
C2
0.01 uF

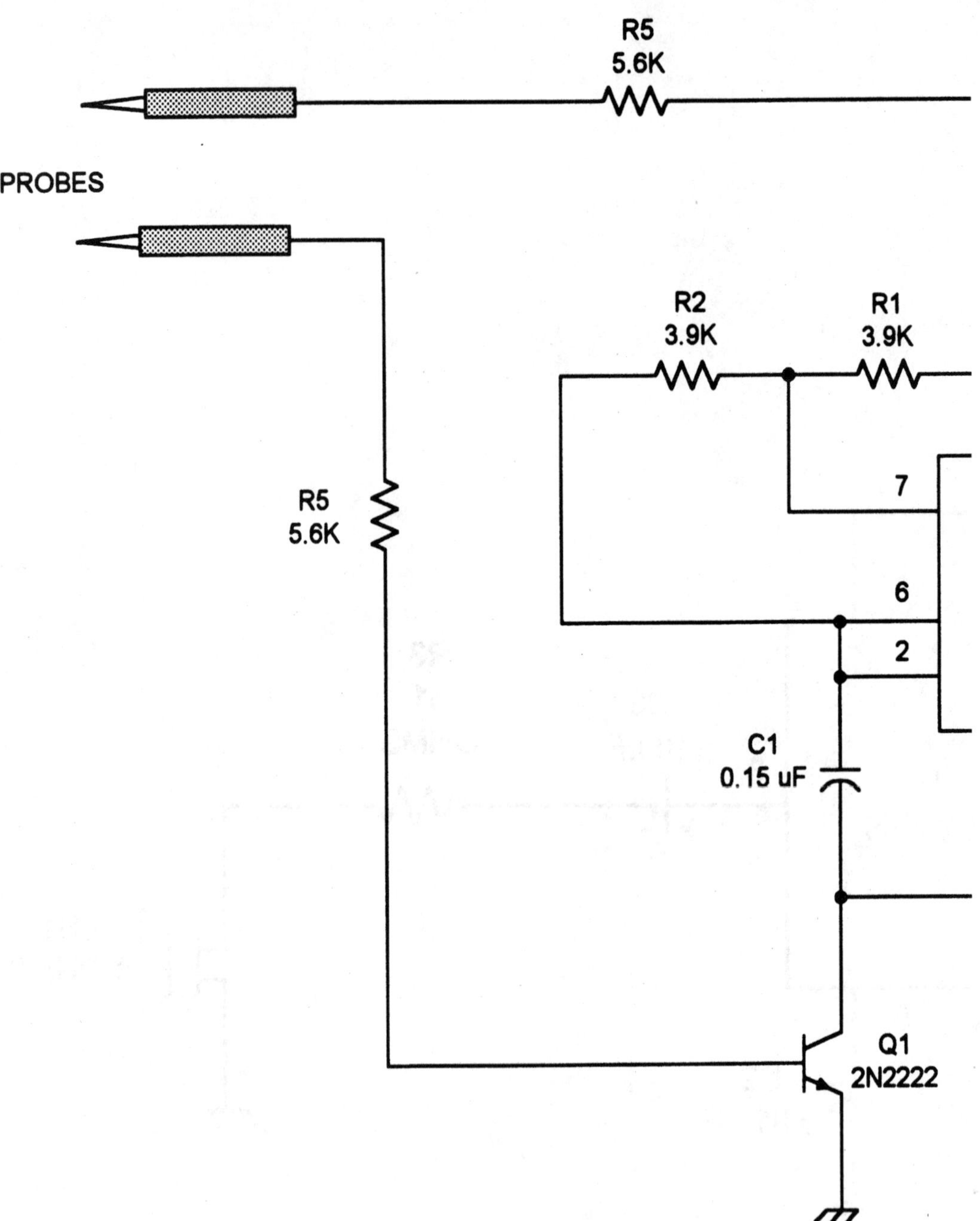

Figure 28-2. Aural continuity tester — Type II.

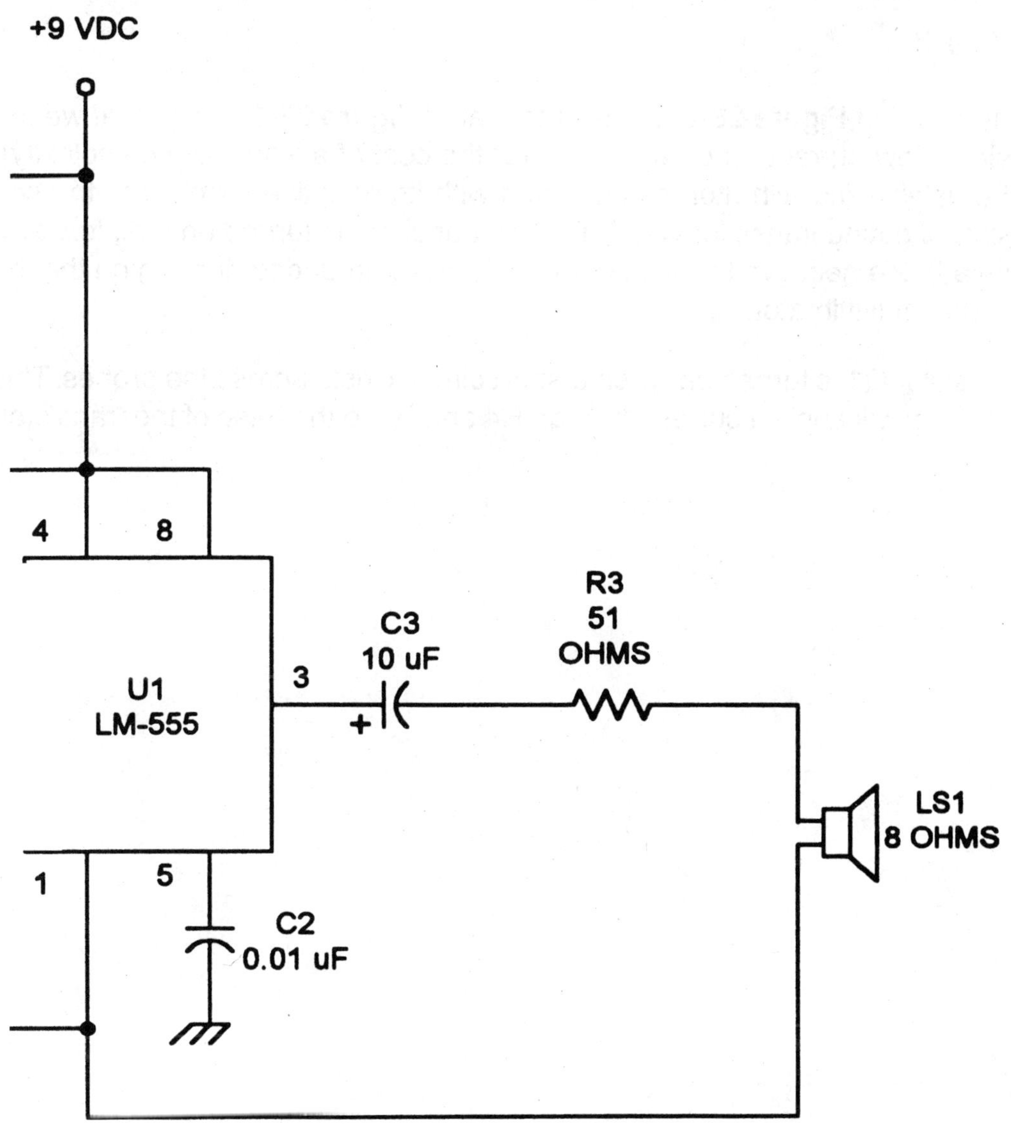
+9 VDC
4
8
U1
LM-555
3
C3
10 uF
+
R3
51
OHMS
LS1
8 OHMS
1
5
C2
0.01 uF

Type II

The circuit in **Figure 28-2** is similar to that of **Figure 28-1** except that we are using a low-current probe approach (at the cost of a little more complexity). The astable multivibrator is connected with its negative terminal to the collector of gating transistor Q1. When this transistor is turned on (i.e., forward-biased), the negative terminal of the LM-555 is grounded, turning on the the astable multivibrator.

Transistor Q1 is turned on when a short circuit exists across the probes. This condition will apply a current through R4 and R5 to the base of the transistor.

Chapter 29

100-kHz Crystal Calibrator

Shortwave listeners and ham radio operators sometimes use 100-khz crystal calibrators to check the calibration of receiver dials (especially the old-fashioned analog type). **Figure 29-1** shows a 100-kHz crystal calibrator circuit based on the LM-555 timer IC. The frequency of the astable multivibrator is determined by a piezoelectric crystal. In this case, we have selected a 100-kHz crystal. This frequency allows us to place harmonic markers at frequencies throughout the high frequency (HF) shortwave bands, and even into the low VHF bands.

The output of the circuit is a square wave, so is rich in harmonics. The harmonic content can be improved somewhat by placing a transistor amplifier between the output of the LM-555 and the load. The amplifier is designed to have a diode rectifier in the collector load circuit.

The crystal calibrator circuit will work throughout the HF region, but is limited by the fact that the marker signals only occur at 100-kHz intervals. If you use frequency dividers, then you can provide 50-kHz, 25-kHz, 10-kHz, 5-kHz and other outputs derived from the 100-kHz signal.

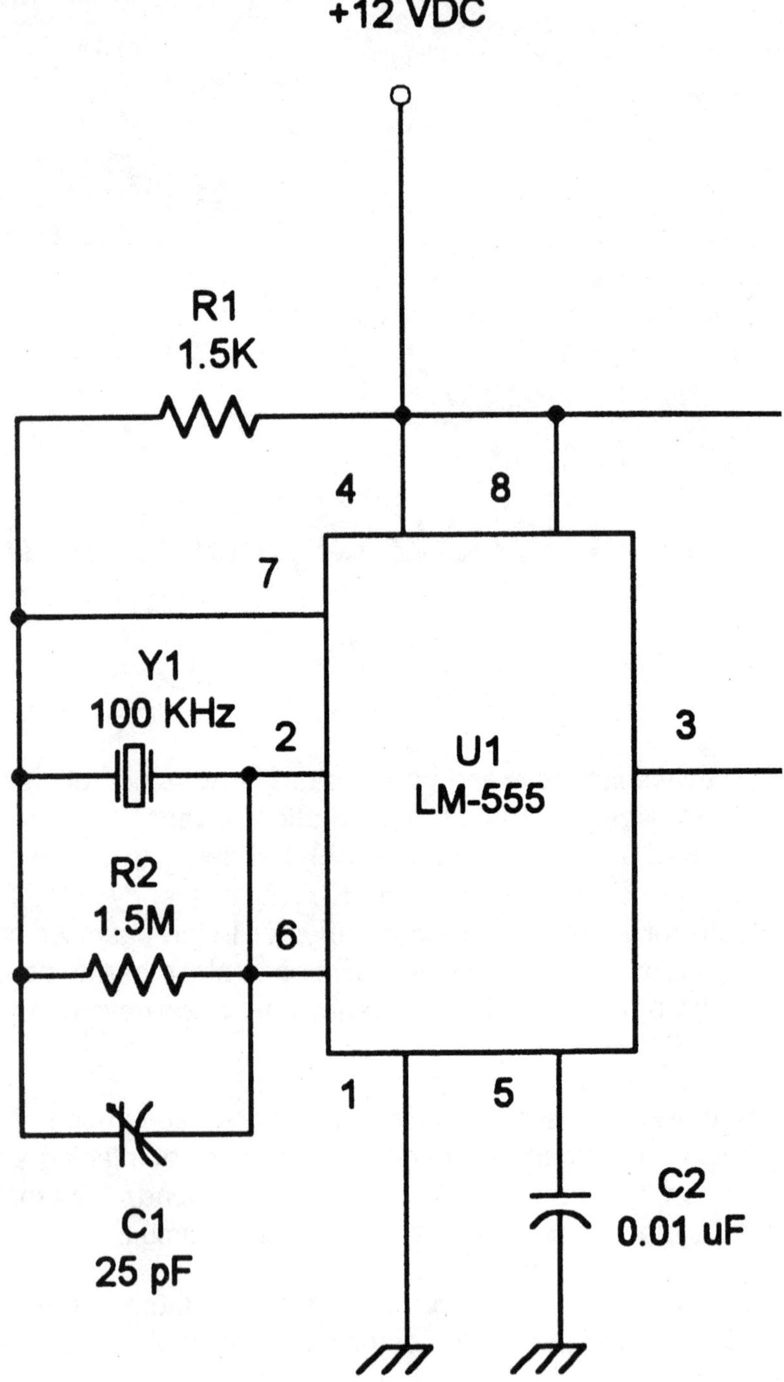

Figure 29-1. 100-khz crystal calibrator.

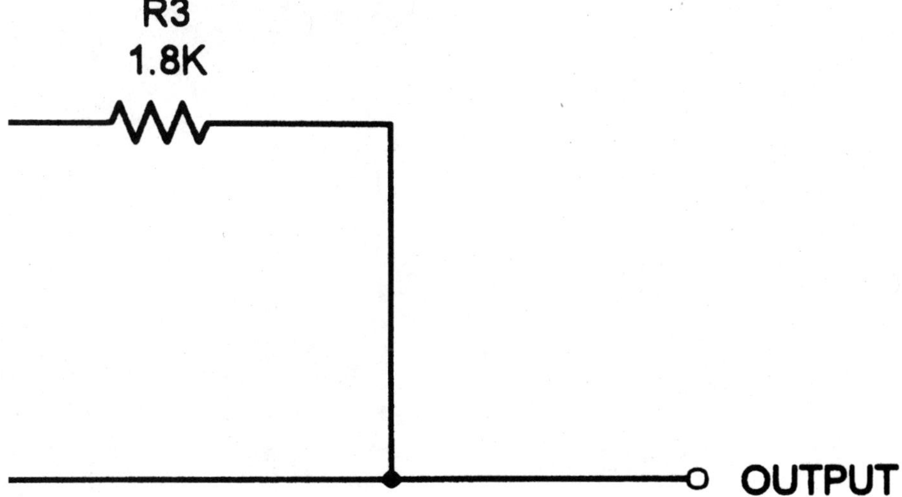
R3
1.8K
OUTPUT

Chapter 30

Pulse Catcher Circuit

Many digital circuits produce very short duration, high speed transient pulses that are difficult to catch on an oscilloscope (especially lower cost or non-digital scopes). We sometimes try to catch transient pulses on a storage scope or with an oscilloscope camera. But some pulses are simply too fast for the writing speeds of either storage scope or scope cameras. The circuit in **Figure 30-1** solves that problem: it is a pulse catcher.

The pulse is coupled to the circuit through a low-value disk ceramic capacitor (C1, 33 pF) to the input of a CMOS 4050 noninverting buffer. The output of the 4050 is applied to the clock (CLK) input of a CMOS 4013 Type-D flip-flop. Since the D input of this stage is wired permanently HIGH, the Q output will go HIGH when the pulse is received from the 4050. A Reset switch (S1) will cause the normally LOW clear (CLR) input to go HIGH when we want to reset the circuit.

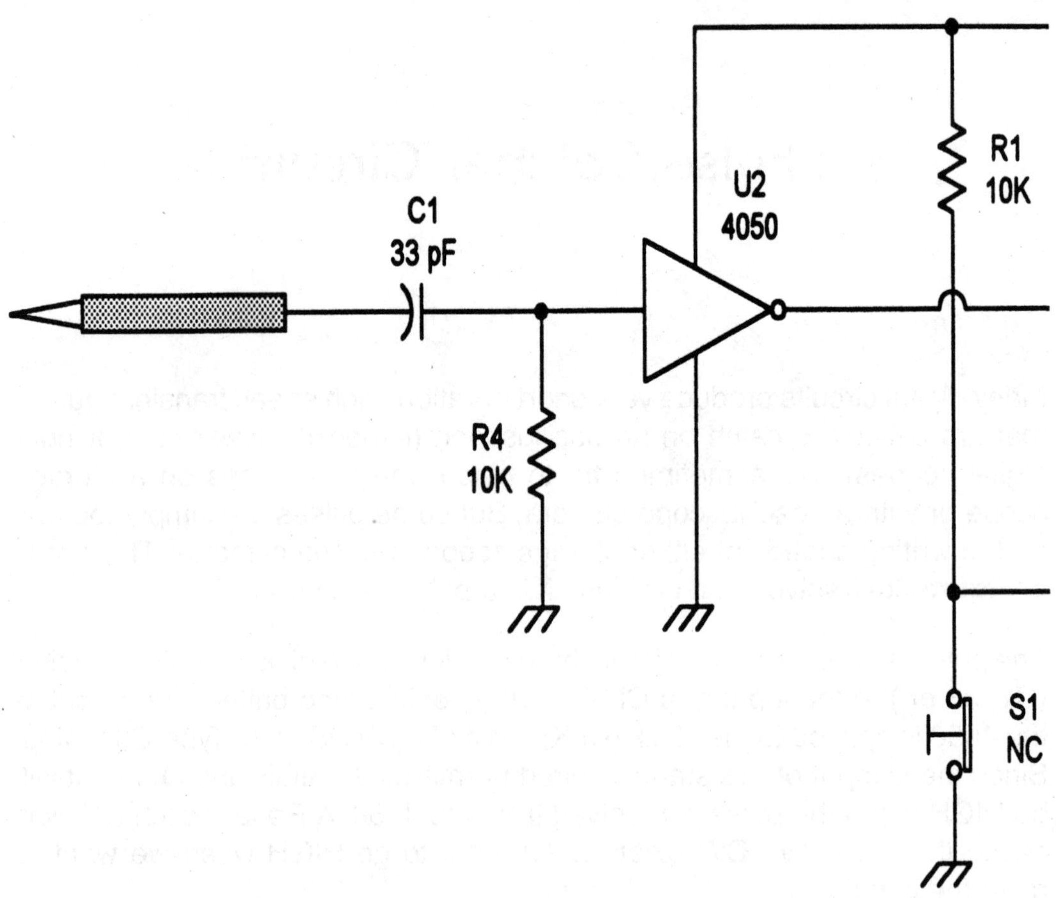

Figure 30-1. Pulse catcher.

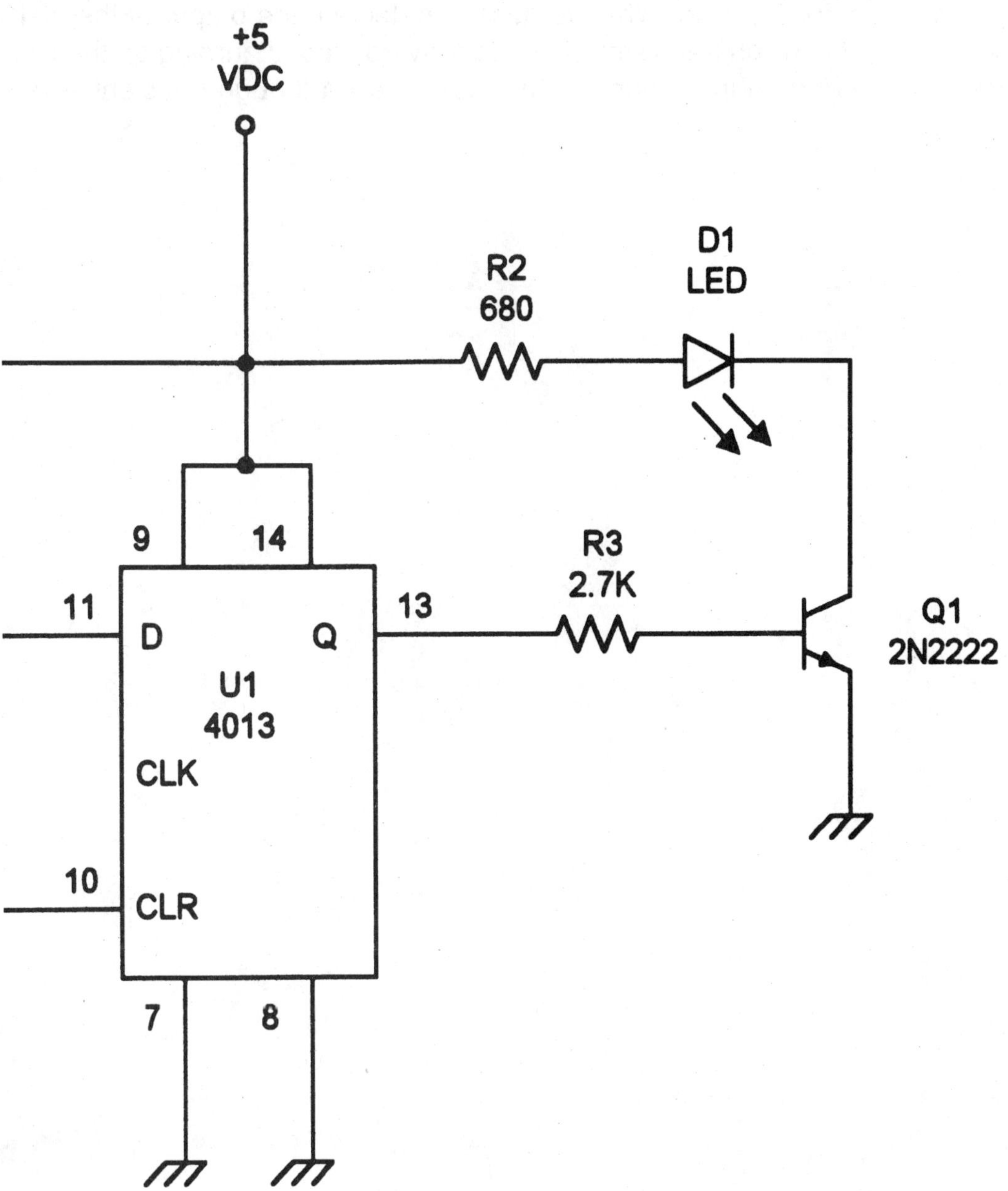
+5
VDC
R2
680
D1
LED
9
14
R3
2.7K
11
13
D
Q
Q1
2N2222
U1
4013
CLK
10
CLR
7
8

This circuit is used in testing digital logic circuits. Press the Reset switch, clearing the 4013 device. When a pulse is detected, the output of the 4013 will go HIGH causing transistor Q1 to be forward biased, turning on the LED (D1), and giving an indication that a pulse — even though transient — existed.

Index

A

B

C

D

E

F

G

H

I

J

L

M

N

O

P

Q

R

S

T

U

V

W

X

Z

Coming in July, 1997!

Electronic Circuit Guidebook, Volume 3: Op-Amps

The operational amplifier (op amp) is the most commonly used linear IC amplifier in the world. The range of applications for the op amp is truly awesome — it has become a mainstay of audio, communications, TV, broadcasting, instrumentation, control and measurement circuits.

Electronic Circuit Guidebook, Volume 3: Op Amps is designed to give you some insight into how practical linear IC amplifiers work in actual, practical real-life circuits. Because of their widespread popularity, operational amplifiers (op amps) figure heavily in this book, though other types of amplifiers are not overlooked. ***Electronic Circuit Guidebook, Volume 3: Op Amps*** allows you to design and configure your own circuits, and is intended to be a practical workshop aid.

Electronic Circuit Guidebook, Volume 3: Op-Amps
340 pages • Paperback • 7-3/8 x 9-1/4"
ISBN: 0-7906-1131-5 • Sams: 61131
$24.95 • July 1997

CALL 1-800-428-7267 TODAY FOR THE NAME OF YOUR NEAREST PROMPT PUBLICATIONS DISTRIBUTOR